纺织服装类"十四五"部委级规划教材
校企合作双语教材

Pattern
Design and CAD

服装纸样设计与CAD

（中英文对照）

Chief Editor: Cao Qiong
主编：曹 琼
Assoiate Editor: Wu Yonghong、Ba Guiling、Zhu Meifang
副主编：吴永红、巴桂玲、竺梅芳
English Proof Reader: Shay McCarthy–Symes
英文审读：Shay McCarthy–Symes

东华大学 出版社 · 上海

图书在版编目 (CIP) 数据

服装纸样设计与 CAD ：汉英对照 / 曹琼主编；吴永红，巴桂玲，竺梅芳副主编 . -- 上海 ： 东华大学出版社，2022.5

ISBN 978-7-5669-2016-4

Ⅰ．①服… Ⅱ．①曹… ②吴… ③巴… ④竺… Ⅲ．①服装设计－纸样设计－计算机辅助设计－AutoCAD 软件－汉、英 Ⅳ．① TS941.26

中国版本图书馆 CIP 数据核字（2021）第 247940 号

责任编辑：谭 英
封面设计：Marquis

服装纸样设计与 CAD（汉英对照）

Fuzhuang Zhiyang Sheji yu CAD

主 编 曹琼
副 主 编 吴永红、巴桂玲、竺梅芳
英文审读 Shay McCarthy–Symes

东华大学出版社出版

上海市延安西路 1882 号

邮政编码：200051 电话：（021）62193056

出版社网址： http://www.dhupress.dhu.edu.cn

天猫旗舰店： http://www.dhdx.tmall.com

印刷：上海盛通时代印刷有限公司

开本：787 mm×1092 mm 1/16 印张：11.5 字数：405 千字

2022 年 5 月第 1 版 2022 年 5 月第 1 次印刷

ISBN 978-7-5669-2016-4

定价：47.00 元

Preface

Pattern design is not only an art but also a science. As an art, it pursues beauty; As a science, it is required to fit the human body, and be comfortable to wear. Therefore, it is not a simple cutting, but a comprehensive subject involving ergonomics, aesthetics, fashion design, fashion technology design and clothing fabric and accessories knowledge. Because of its importance in clothing, clothing pattern design course has always been the core course of clothing design major and clothing engineering major in colleges or universities. Women's wear styles are ever-changing and structural changes are much more abundant than men's wear. Therefore, clothing pattern design courses mainly teach women's fashion changes.

This book is one of the planning textbooks for the construction of high quality specialty in Zhejiang Fashion Institute of Technology. This book is only a basic part about women's wear, and then I plan to write a ready-to-wear part. As a basic part, this book focuses on the basic knowledge of clothing pattern design, basic pattern design, clothing body changes, collar pattern design, sleeve pattern design and comprehensive training.The style diagram of this book is mostly drawn according to the sample taken after the sample is photographed on the mannequin , which ensures the length and width ratio of the style diagram is not out of balance. With the principle of practicality and adequacy, this book simplifies a wide range of theoretical knowledge and only selects the applicable theoretical knowledge for explanation. Many examples of structural changes in the book are representative styles, focusing on qualitative analysis and teaching pattern change methods. In the comprehensive training part, the contents of the garment body, collar and sleeve, which have been learned before, are combined and applied with examples to train the students' strain capacity.

This textbook is chiefly edited by Cao Qiong from Zhejiang Fashion Institute of Technology, who is responsible for the writing of the main chapters and the drafting and modifying of the book. The deputy chief editors of the book are Wu Yonghong from Nanchang University, Ba Guiling and Zhu Meifang from Zhejiang Fashion Institute of Technology. The content of the book is divided into six chapters. Unit 1 and 2 of Chapter 1, Chapter 3, and Unit 1 of Chapter 6 were written by Cao Qiong ; Unit 3 and 4 of Chapter 1, and Chapter 4 were written by Ba Guiling; Chapter 2 was written by Pu Haiyan of Jiaxing University; Chapter 5, Units 2 and 3 of Chapter 6 were written by Wu Yonghong ; The CAD content of the last unit of each chapter was compiled by Zhu Meifang.

This textbook is also a school-enterprise cooperation textbook. China Modasoft co., LTD provided "Zhizunbaofang CAD" for free use by our students. The general manager of the company Huang Yanjun also provided the sponsorship for the publication of this book. As a result, the CAD parts of this book are all written using "Zhizunbaofang CAD". Here, I would like to express my special thanks to Mr. Huang Yanjun. I also want to express my special thanks to Shay McCarthy-Symes, an international student who came to our school from Manchester School of Art. As an English proof reader, she read through the book and helped to proofread the English part of the book, making it a professional clothing book that foreigners can understand. I would like to express my special thanks to her!

In terms of drawing in this book, it received some help from Wen Ying, who is a teacher of Zhejiang Fashion Institute of Technology, and some students such as Xu Jingxian, Qiu Kaiwen, Chen Yang, etc. Here, I would like to express my gratitude to them!

Due to our limited time and limited level, it is unavoidable that there are shortcomings in this book. Welcome criticism from the experts, peers and readers. We will be very grateful!

Cao Qiong

March 20, 2022

前 言

    服装纸样设计既是一门艺术，更是一门科学。作为一门艺术，它追求的是美观；作为一门科学，它要求适合人体，穿着舒适。因此，它不是简单的裁剪，而是一门涉及人体工学、美学、服装款式设计与工艺设计，以及服装面、辅料知识的一门综合性学科。因其在服装上的重要性，服装纸样设计课程也历来是各大专院校服装设计与服装工程专业的专业核心课程。女装款式千变万化，其结构变化比男装丰富，因此，本书中服装纸样设计以讲授女装变化为主。

    本书是浙江纺织服装职业技术学院的优质专业建设的规划教材之一。本书只是女装基础篇，之后计划编写成衣篇。作为基础篇，本书着重编写了服装纸样设计的基础知识、基础纸样、服装衣身变化、领子纸样设计、袖子纸样设计及综合训练等内容。本书的款式图大多是根据样衣在人台上试样并拍照后描画出来的，保证了款式图的长宽比例不失调。以实用、够用为原则，本书精简了繁多的理论知识，只选取了适用的理论知识进行讲解。书中的许多结构变化的例子都是代表性的款式，讲解偏重定性分析，以讲授纸样变化方法为主。综合训练部分主要以实例的形式，将前面学过的衣身、领子、袖子部分的内容加以组合应用，以训练学生的应变能力。

    本教材由浙江纺织服装职业技术学院的曹琼主编，负责主要章节的编写和全书的统稿与修改；南昌大学的吴永红、浙江纺织服装职业技术学院的巴桂玲和竺梅芳为副主编。全书共分六章。第一章第一、二节，第三章，第六章第一节，由曹琼编写；第一章第三、四节，第四章，由巴桂玲编写；第二章由嘉兴学院浦海燕编写；第五章，第六章第二、三节，由吴永红编写；全书每章最后一节的 CAD 方面内容都由竺梅芳编写。

    本教材还是一本校企合作教材。上海深厚数码科技有限公司免费提供了 CAD 软件"智尊宝纺"给我校学生使用，该公司黄彦军总经理为本书的出版还提供了赞助。因此，本书的 CAD 部分编写都是使用"智尊宝纺"软件进行的。在此，对黄彦军先生表示特别感谢！另还需特别感谢的是从英国 Manchester School of Art 来我校留学的同学 Shay McCarthy-Symes。她通读了本书，帮助校对了本书的英文部分。在此表示感谢！

    本书编写过程中的图片绘制，得到了浙江纺织服装职业技术学院文英老师和许静娴、裘凯闻、陈洋等几位学生的帮助，在此一并表示感谢！

    由于时间、水平有限，书中难免有不足之处，欢迎各位专家、同行和读者批评指正，将不胜感谢！

曹琼

2022 年 3 月 20

Contents / 目录

Chapter 1 Clothing and the body
第一章 服装与人体

Pattern cutting is a means of achieving a shape around the body. It is very important and necessary that students on clothing courses understand the body form. Pattern making is a creative skill. It is simple and exciting if the basic principles are learned thoroughly and the student is perceptive to the slight changes made to fashion shapes.

纸样裁剪是塑造人体造型的一种手段。学习服装课程的学生了解些关于人体的知识是非常重要和必要的。纸样制作是一种创造性的技能。如果学生彻底地学习了纸样制作的基本原理并且能敏锐感受到时装上造型的细微变化，纸样制作就会变得很简单并且令人激动。

Unit 1 Critical points and lines on the body and in clothing
第一节 人体与服装上的关键结构点与线

一、Critical structure points on the body and in clothing/ 人体与服装上关键结构点

1. Top point of head / 头顶点

It is the highest point of the head that is located on the center-line of the body and is a reference point for measuring height when a person stands correctly in an upright posture.

以正确直立姿势站立时，该点是头部最高点，位于人体中心线上，它是测量人体身高时的基准点。

2. Front neck point / 前颈点

It is the front center point of the neck curve. It is an important reference point for defining the front neck-line of clothing.

该点是颈根曲线的前中心点，也是衣身前领圈的重要参考点。

3. Side neck point / 侧颈点

It is on the neck curve of the body, slightly rearward at the midpoint of neck thickness when viewed from the side of the neck. It's not easy to be found because there is no obvious endpoint bone here.

该点是在颈根的曲线上，从侧面看在前后颈厚之中央稍微偏后的位置。此点不是人体骨骼端点，所以不易确定。

4. Back neck point / 后颈点

It becomes protruding and easy to be found when bending neck forward; it is a reference point for measuring back length.

当颈部向前弯曲时，该点就突出，较易找到，是测量背长的基准点。

5. Shoulder point / 肩点或肩峰点

It is a turning point connecting the shoulder with the arm; it is a reference point to allow the sleeve to match up with the armhole and to measure shoulder width and sleeve length.

肩点是连接肩与手臂的转折点，是衣袖缝合对位的基准点，也是量取肩宽和袖长的基准点。

6. Front armhole point / 前腋点

It is the start point at the joint of the arm and the front torso of the body when the arm hangs down, and it is a reference point for measuring chest width.

放下手臂时，该点是手臂与前躯干部在腋下结合处之起点，是测量胸宽的基准点。

7. Back armhole point / 后腋点

It is the start point at the joint of the arm and the back torso when the arm hangs down, and it is a reference point for measuring back width.

放下手臂时，该点是手臂与后躯干部在腋下结合处之起点，是测量背宽的基准点。

8. Bust point / 乳突点或胸点

It is the most protruding point on the chest and one of the most important basic points when creating patterns.

该点是胸部最高的地方，也是服装纸样上最重要的基准点之一。

9. Elbow point / 肘点

It is the most protruding point from the upper end of the ulna, and is clearly protruding when the upper limb is naturally bent and is a reference point for measuring the length of the upper arm.

该点是尺骨上端向外最突出之点，上肢自然弯曲时该点很明显地突出，是测量上臂长的基准点。

10. Wrist point / 手腕点

It is the top point at the lower end of the radius and is a reference point for measuring sleeve length.

该点是桡骨下端最尖端点，是测量袖长的基准点。

11. Knee cap point / 髌骨点

It is in the center of the knee bone and located at the front of the knee joint.
该点位于膝关节前部的膝盖骨之中央。

12. Ankle point / 外踝点

It is the protruding point on the outer side of the ankle and is a reference point for measuring the trousers length.

该点是脚腕外侧踝骨的突出点，是测量裤长的基准点。

These critical points are shown in Fig.1-1-1.

这些关键点如图 1-1-1 所示。

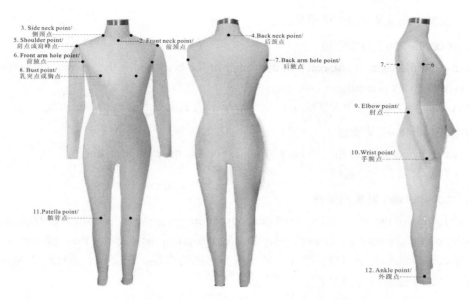

Fig.1-1-1 critical points on the body and in clothing
图 1-1-1 人体和服装上的关键点

二、Critical structure lines on the body and in clothing / 人体与服装上关键结构线

1. Front centerline / 前中心线

It is located in the center of the body when viewed from the front and is the reference line for setting the positions of buttonholes and buttons.

从正面看人体该线位于人体的中央位置，是服装上设置纽扣眼和纽扣的基准线。

2. Neck curve / 颈根曲线

It is the juction of the neck and torso of the body, and it is an irregular curve. It is usually considered to be a smooth round line in clothing and is the reference line for designing the shape of the collar curve.

该线是人体颈部与躯干部的交接线，是一条不规则的弧线。在服装上它通常被当着是一条光滑的圆弧线，是设计领口造型的参照线。

3. Shoulder line / 肩线或肩斜线

It is formed by connecting the side neck point to the shoulder point, and it is an important structure line connecting the front and rear pieces of clothing.

该线由侧颈点与肩点连接而成，是服装上连接前片与后片的重要结构线。

4. Chest width line / 胸宽线

It is formed by connecting the left front armhole point to the right one.

连接左前腋点到右前腋点形成胸宽线。

5. Bust point line / 乳峰线

It is a line formed by circling around the fullest part of the bust through the bust point (BP); it is the reference line for measuring the bust of the body, although it's not the bust line in clothing.

该线为人体上经过胸点（BP）的绕胸部最丰满处一圈形成的线条。它是测量人体胸围尺寸的基准线，虽然它不是服装上的胸围线。

6. Back centerline / 后中心线

Viewed from the back, it is located in the center of the body and is the reference line for setting the position of the back centerline on the garment.

从背面看人体，该线位于人体的中央位置，是服装上设置后中心线的基准线。

7. Back width line / 背宽线

It is formed by connecting the left back armhole point to the right back one.

连接左右后腋点形成背宽线。

8. Armhole curve / 臂根围弧线

It is the juction of the arms and torso of the body, an irregular curve. It is often considered to be a smooth curve on the garment and is a reference line for designing the shape of the armhole.

该线是人体手臂与躯干部的交接线，是一条不规则的弧线。在服装上通常被认为是一条光滑的弧线，是设计袖窿造型的参照线。

9. Waist line / 腰围线

It is formed by circling around the narrowest part of the torso (e.g the waist) and is the reference line for measuring the waist of the body.

该线是围绕腰部最细处一圈的线条，是测量人体腰围尺寸的基准线。

10. Upper arm line / 上臂围线

It is formed around the fullest part of the upper arm and is the reference line for measuring the circumference of the upper arm.

该线是围绕人体上臂最丰满处一圈形成的线条，是测量人体上臂围尺寸的基准线。

11. Elbow line / 肘围线

It is formed around the elbow joint and is the reference line for measuring the circumference of the elbow.

该线是围绕人体肘关节一圈形成的线条，它是测量人体肘围尺寸的基准线。

12. Wrist line / 腕围线

It is formed around the wrist joint and is the reference line for measuring the circumference of the wrist.

该线是围绕人体腕关节一圈形成的线条，它是测量人体腕围尺寸的基准线。

13. Hip line / 臀围线

It is formed around the fullest part of the hip and is the reference line for measuring the circumference of the hip.

该线是围绕臀部最丰满处一圈形成的线条，它是测量人体臀围尺寸的基准线。

14. Crutch depth line / 横裆线

It is the line across the lower part of the hip.
该线位于两条大腿的分叉处。

15. Knee line / 膝围线或髌骨线

It is formed around the knee bone and is the reference line for measuring the circumference of the knee. It is also the reference line for designing the position of the knee line in trousers.

该线是绕膝盖骨一圈形成的线条，它是测量人体膝盖围度尺寸的基准线，也是设计裤子中档线位置的参照线。

16. Ankle line / 脚踝线

It is formed around the ankle joint and is the reference line for measuring the circumference of the ankle, and it is the reference line for designing the opening position of trouser legs.

该线是绕踝关节一圈形成的线条，它是测量人体踝关节围度尺寸的基准线，也是设计裤子脚口线位置的参照线。

These critical lines are shown in Fig.1-1-2

这些人体上关键的线如图 1-1-2 所示。

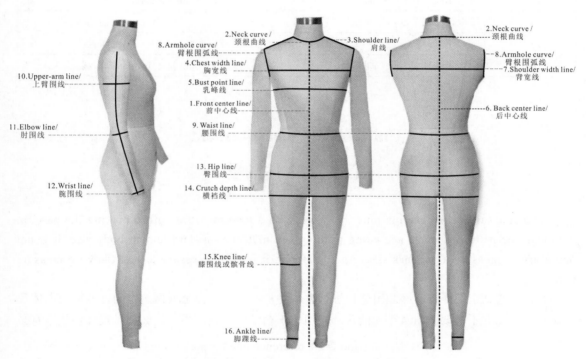

Fig.1-1-2 The critical structure lines of the body and clothing
图 1-1-2 人体和服装上的关键线

The critical structure lines of the body and clothing are shown in table1-1-1.

人体和服装上关键线的名称如表 1-1-1 所示。

Table 1-1-1 The critical structure lines of the body and clothing
表 1-1-1 人体和服装上的关键线

number （序号）	The critical structure lines on the body （人体上的关键线）	The critical structure lines in clothing （服装上的关键线）
1	Front centerline（前中心线）	Front centerline（前中心线）
2	Neck curve （颈根曲线）	Collar curve（领口弧线）

（续表）

3	Shoulder line（肩线）	Shoulder line（肩线）
4	Chest width line（胸宽线）	Chest width line（胸宽线）
5	Bust point line（乳峰线）	Bust line（胸围线）
6	Back centerline（后中心线）	Back centerline（后中心线）
7	Back width line（背宽线）	Back width line（背宽线）
8	Armhole curve（臂根围弧线）	Armhole curve（袖窿弧线）
9	Waist line（腰围线）	Waist line（腰围线）
10	Upper arm line（上臂围线）	Sleeve width line（袖宽线）
11	Elbow line（肘围线）	Elbow line（袖肘线）
12	Wrist line（腕围线）	Sleeve opening line（袖口线）
13	Hip line（臀围线）	Hip line（臀围线）
14	Crutch depth line（横裆线）	Crutch depth line（横裆线）
15	Knee line（膝围线或髌骨线）	Knee line（中裆线或髌骨线）
16	Ankle line（脚踝线）	Leg opening line（脚口线）

Unit 2 Body measurement

第二节 人体测量

As a rule, all measurements must be taken with a tape measure before making the pattern. Although the pattern is made according to the given industry standard human body size, it is not necessary to measure a person's size. As a beginner, it is still necessary to learn how to measure correctly.

通常，在制作纸样之前必须用软尺进行人体测量。尽管纸样是根据给定的行业标准人体尺寸制作的，而并不需要测量某个人的尺寸。但对于初学者来说，学习怎么测量尺寸仍然是有必要的。

一、Procedure for taking measurements / 测量尺寸的程序

First measure the length, then measure the width and the circumference; first measure the body, then the limbs; first measure the upper body, then the lower body.

先量长度，再量宽度和围度；先量正身，再量四肢；先量上半身，再量下半身。

二、Taking measurements / 测量尺寸

Measure according to the following instructions and Fig. 1-2-1, Fig. 1-2-2.

根据下面的操作指南和图 1-2-1、图 1-2-2 进行测量。

1. Body height / 身高

Measure the vertical distance from the top of your head to the floor.

测量从头顶到地面的垂直距离。

2. Front side neck point to waist / 前腰节长

Measure the length from the side neck point via the bust point and then vertical to the waistline.

自侧颈点开始，经 BP 点，再垂直至腰围线的长度。

3. Nape to waist / 背长

Measure from the back neck point to the waistline along the back centerline.

在后中从后颈点量到腰围线处。

4. Shoulder length / 肩线长

Measure from the side neck point to the shoulder point.

从侧颈点量到肩点。

5. Chest width / 胸宽

About 6 cm down from the front neck point along the front centerline, measure from the left front armhole point to the right front armhole point.

过前中颈点向下约 6cm 处，从左前腋点量到右前腋点。

6. Shoulder width / 肩宽

Measure from the left shoulder point to the right shoulder point on the back of the torso.

在后背上从左肩点量到右肩点。

7. Back width / 背宽

Roughly 12 cm down from the neck bone along the back centerline, measure the distance from the left back armhole point to the right back armhole point.

过后中颈椎骨向下约 12cm 处，从左后腋点量到右后腋点。

8. Neck size / 颈围尺寸

Measure around the neck root through the front neck point.

经过前颈点围绕颈根部一圈测量。

9. Bust / 胸围

Measure around the fullest part of the chest through the bust point, and do not let the tape drop off the back torso.

经过乳点围绕胸部最丰满处一圈进行测量，不要让皮尺在背部落下。

10. Waist / 腰围

Measure around the narrowest part of waist; make sure that it is comfortable.

围绕腰部最细处一圈进行测量，要确保腰部舒适。

11. Hip / 臀围

At the position 18 to 20 cm down from the waistline, measure around the fullest part of the hip.

在腰围线下 18 ～ 20cm 处，围绕臀部最丰满处一圈进行测量。

12. Armhole / 臂根围

Measure around the junction of the human arm and torso.

围绕人体手臂与躯干部的交接处一圈进行测量。

13. Sleeve length / 臂长

Let the arms hang down naturally. Measure the distance from the shoulder point through the elbow to the wrist point.

让手臂部自然下垂。从肩点经过肘部量到手腕点。

14. Upper arm / 上臂围

With the arm bent, measure around the largest circumference of the upper arm.

手臂弯曲，围绕上臂围度最大的地方一圈进行测量。

15. Elbow / 肘围

Measure around the elbow joint through the elbow bone.

经过肘骨绕肘关节一圈进行测量。

16. Wrist / 腕围

Measure around the wrist joint through the wrist point loosely, leaving room for ease.

经过手腕点绕腕关节一圈进行测量，留点松量。

17. Palm / 掌围

Measure around the widest part of the palm.

围绕手掌最宽部位一周进行测量。

18. Waist to floor / 下身长

Measure the distance from the waist line to the floor along the back centerline.

在后中从腰围线量到地面。

19. Waist to hip / 臀长

Measure the distance from the waist line to the hip line.

从腰围线量到臀围线。

20. Body rise / 立裆

Have the subject sit on a hard chair. On the side of the body, measure the distance from the waist line to the seat surface of the chair.

让被测者坐在一把硬椅子上，在侧面从腰围线量到椅子面。

21. Waist to knee / 腰围至膝围长

Measure the distance from the waist line to the knee line.

测量从腰围线到膝围线的距离。

22. Ankle / 脚踝围

Measure around the ankle joint through the ankle bone.

经过踝骨绕踝关节一圈进行测量。

23. Head size / 头围

Measure in a circle around the head in the middle of the forehead.

前额中部绕头部一圈进行测量。

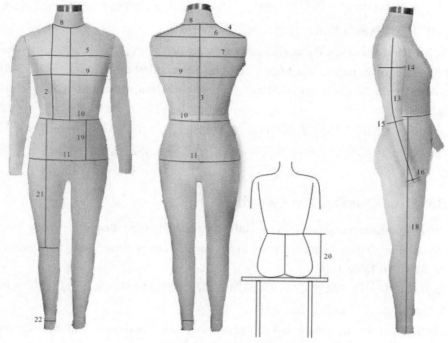

Fig.1–2–1 Body measurements
图 1–2–1 人体测量

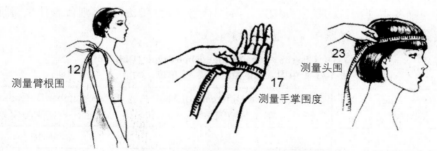

Fig.1–2–2 Part of body measurements
图 1–2–2 部分人体测量

Unit 3 Clothing size

第三节 服装号型

一、Definition of clothing size / 号型定义

Clothing sizes are called "Hao" and "Xing" in Chinese. "Hao" refers to the height of the human body, which is the basis for the design of clothing length. "Xing" refers to the net bust or the net waist of the human body, which is the basis for the design of clothing girth.

衣服尺码在汉语中称为"号"和"型"。"号"指人体的身高，是设计服装长度的依据。"型"指人体的净体胸围或腰围，是设计服装围度的依据。

Height, Bust and Waist are the basic parts of the human body. Using the sizes of these parts to calculate the sizes of other parts of the human body has the smallest error. In the GB1335–97 standard, it is determined that the height is named "Hao", and the body bust, waist and body type classification code named "Xing".

身高、胸围和腰围是人体的基本部位，用这些部位的尺寸来推算其他各部位的尺寸，误差最小。在 GB1335—97 标准中确定将身高命名为"号"，人体胸围和人体腰围及体型分类代号命名为"型"。

二、Body type Classification / 体型分类

Body type classification is based on the difference in body type, that is, net body bust minus net body waist. According to this, the human body in our country can be divided into four body types: Y, A, B and C. As shown in Table 1-3-1.

体型分类是根据人体的胸腰差即净体胸围减去净体腰围的差数来划分。我国人体可分为四种体型，即 Y、A、B、C。如表 1-3-1 所示。

Clothing Size governs the length and circumference of each main part, and the body type code (Y, A, B, C) represents the body characteristic, therefore the elements of garment size are height, net bust/net waist, and body type code.

服装号型控制各主要部位的长度和围度，体型代号（Y、A、B、C）则代表体型特征，因此服装号型的要素为：身高、净胸围 / 净腰围和体型代号。

Table 1-3-1 Classification of Chinese body type (Unit: cm)
表 1-3-1 中国人体体型的分类　（单位：cm）

Body Type Code （体型分类代号）	Thoracolumbar Difference of Men （男子胸腰围差）	Thoracolumbar Difference of Women（女子胸腰围差）
Y	22 ~ 17	24 ~ 19
A	16 ~ 12	18 ~ 14
B	11 ~ 7	13 ~ 9
C	6 ~ 2	8 ~ 4

三、Size mark / 号型标志

Size mark is the code of clothing size specification. Garment size must be marked on the garment such as "Hao /Xing and Body Type Code". For example, 160/84A means that the height is 160 cm, the net bust is 84 cm, and the body code is A.

号型标志是服装号型规格的代号。成品服装上必须标明：号 / 型，后接体型分类代号。例如，160/84A 表示身高为 160 cm，净胸围为 84 cm，体型分类代号为 A。

四、Medium / 中间体

The Medium human body type size is the mean of a large amount of measured human body data. It reflects the average level of the height, bust, waist, and other parts of all kinds of body types of adult men and women in our country, and has certain representativeness. The middle human type is showed in table 1-3-2.

中间体是根据大量实测的人体数据计算求出的均值，它反映了我国男女成人各类体型的身高、胸围、腰围等部位的平均水平，具有一定的代表性。中间体的设置参见表 1-3-2 所示。

Table 1-3-2 The Medium Human Body Type Size (Unit: cm)
表 1-3-2 男女各体型分类的中间体设置 （单位：cm）

Body Type Code（体型代号）		Y	A	B	C
Man（男子）	Height（身高）	170	170	170	170
	Bust（胸围）	88	88	92	96
	Waist（腰围）	70	74	84	92
	Hip（臀围）	90	90	95	97
Women（女子）	Height（身高）	160	160	160	160
	Bust（胸围）	84	84	88	88
	Waist（腰围）	64	68	78	82
	Hip（臀围）	90	90	96	96

五、Size systems / 号型系列

Size systems refer to the regular permutations and combinations of fashion size. At 5 cm intervals in all sizes of Height in the standard, Men standard measurements are divided into 7 levels: 155 cm, 160 cm, 165 cm, 170 cm, 175 cm, 180 cm and 185 cm; Women standard measurements are also divided into 7 levels: 145 cm, 150 cm, 155 cm, 160 cm, 165 cm, 170 cm and 175 cm. And at 4 cm or 2 cm intervals in all sizes of Bust or Waist respectively, they combine into the 5-4 Size Systems and 5-2 Size Systems. Generally, 5-4 Size Systems are used in tops and 5-2 Size Systems are used in bottoms. These systems are showed in Table 1-3-3 and Table 1-3-4.

号型系列是指将人体的号和型进行有规则的分档排列与组合。在标准中身高规定以 5 cm间隔分档，把男子标准分成 7 档，即从 155cm、160cm、165cm、170cm、175cm、180cm 到 185cm；把女子标准也分成 7 档，即从 145cm、150cm、155cm、160cm、165cm、170cm 到 175cm。加上胸围和腰围分别以 4 cm和 2 cm间隔分档，它们组合成了 5·4 系列和 5·2 系列，上装一般多采用 5·4 系列，下装多采用 5·2 系列。男、女装号型系列分别如表 1-3-3、1-3-4 所示。

Table 1-3-3 5-4A, 5-2A Size Systems for Men (Unit: cm)
表 1-3-3 男装 5·4A、5·2A 号型系列 （单位 :cm）

	A															
胸围	身高															
	155		160		165		170		175		180		185			
	腰围															
72				56	58	60	56	58	60							

（续表）

76	60	62	64	60	62	64	60	62	64	60	62	64									
80	64	66	68	64	66	68	64	66	68	64	66	68	64	66	68						
84	68	70	72	68	70	72	68	70	72	68	70	72	68	70	72	68	70	72			
88	72	74	76	72	74	76	72	74	76	72	74	76	72	74	76	72	74	76	72	74	76
92				76	78	80	76	78	80	76	78	80	76	78	80	76	78	80	76	78	80
96							80	82	84	80	82	84	80	82	84	80	82	84	80	82	84
100										84	86	88	84	86	88	84	86	88	84	86	88

Table 1-3-4 5·4, 5·2A Size Systems for Women   (Unit: cm)

表 1–3–4 女装 5・4、5・2A 号型系列   （单位：cm）

胸围	A / 身高 / 腰围																				
	145			150			155			160			165			170			175		
72				54	56	58	54	56	58	54	56	58									
76	58	60	62	58	60	62	58	60	62	58	60	62	58	60	62						
80	62	64	66	62	64	66	62	64	66	62	64	66	62	64	66	62	64	66			
84	66	68	70	66	68	70	66	68	70	66	68	70	66	68	70	66	68	70	66	68	70
88	70	72	74	70	72	74	70	72	74	70	72	74	70	72	74	70	72	74	70	72	74
92				74	76	78	74	76	78	74	76	78	74	76	78	74	76	78	74	76	78
96				78	80	82	78	80	82	78	80	82	78	80	82	78	80	82	78	80	82

Unit 4 Knowledge for Making Patterns

第四节 纸样制作知识

一、Methods of pattern making / 纸样设计的方法

The methods used by a pattern maker to create patterns include flat cutting and draping.

样板师制作纸样的方法主要包括平面制图和立体裁剪两大类。

1. Draping / 立体裁剪

Model the garments on the dress stand. The fabric covering the human body or the mannequin, is dealed by folding, gathering, pleating and other techniques to make the main shape of the garment, and flattens into a two-dimensional pattern.

在人体模型上直接塑造服装造型。将覆合在人体或人体模型上的布料，通过折叠、收省、褶裥等手法做成服装主体形态，并展平成二维的布样。

2. Flat cutting / 平面制图

（1）Block patterns / 原型法

A block pattern is a foundation pattern constructed to fit an average body shape. The block patterns include the important parts of the human body and the ease required for the function of block pattern. The pattern maker takes the prototype as the basis of pattern design, and completes the final required sample by adding style lines, pleats, gathers and other techniques.

原型纸样是适用于一般体型的基本纸样。原型纸样中包含基础样板功能性所需的人体重要部位和基本松量。样板师以原型作为纸样设计的基础，通过增加款式线、褶裥、抽褶等手法完成最终需要的样板。

（2）Basic patterns / 基型法

Generally, each company stores many kinds of patterns, so the designer can choose the pattern that is closest to the desired, then copy and modify it to be the finished pattern. Perhaps, it is the most common use of pattern cutting programs by most companies .

一般来说，企业里面都保存有大量的样板，所以设计师可以从中找出最接近设计款式造型的服装纸样作为基型，然后复制并修改基型，作出所需服装款式的纸样。这也许是大多数公司在设计纸样时最常用的方法。

（3）Formulas of proportion / 比例公式制图法

According to the regression relationship between the basic parts of the human body (height, net bust, and net waist) and the others, you can obtain the formulas of proportion of each part. Generally, the detailed length, for example, the clothes length, the sleeve length, and the length of the trousers, can be expressed by the height: Y=ah+b (h=height, a and b are constants); also, the detailed width of tops, for example, the shoulder width, the chest width and the back width, can be expressed by the net bust or bust (on the basis of the net bust, a loose amount is added): Y=aB+b (B=net Bust or Bust, a and b are constants); so the detailed width of bottoms, can be expressed by the waist or hip: Y=aW+b or Y=aH+b (W=Waist, H=Hip, a and b are constants).

根据人体的基本部位（身高、净胸围、净腰围）与其他部位之间的回归关系，求得各部位尺寸用基本部位尺寸表达的比例公式。一般来说，衣长、袖长、裤长等长度尺寸可用身高的比例公式来表达 : Y=ah+b（h 为身高 , a、b 为常数）；肩宽、胸宽、背宽等上装尺寸可用净胸围或胸围（在净胸围基础上加放了松量）的比例公式来表达：Y=aB+b（B 为净胸围或胸围，a、b 为常数）；下装的尺寸可用腰围或臀围的比例公式来表达：Y=aH+b 或 Y=aW+b（H 为臀围，W 为腰围，a、b 为常数）。

（4）Taking personal measurements / 实寸法

The patterns can be drafted to fit individual figures by using personal or personal garments measurements instead of standard ones listed in the size chart.

实寸法是通过测量实际个体或特定服装的尺寸而不是标准体尺寸来制作服装样板的方法。

二、Tools and equipment for making patterns / 绘制纸样的工具及材料

A student should try to acquire a good set of equipment. However, some items are very expensive. The items marked with an asterisk denote those that are not essential immediately.

初学者应该有一套较好的制图工具，虽然有一些工具比较贵。下面有＊标志的为非必需项。

① Working surface / 工作台。

A flat working surface is required. Ideally, it should be 90–92 cm high.

用于结构制图的平台，理想高度为 90 ～ 92cm。

② Strong brown paper / 牛皮纸。

Strong brown paper is used for patterns. Parchment or thin cardboard should be used for pattern blocks used frequently.

绘制纸样所用的是牛皮纸。频繁使用的纸样应该绘制在牛皮纸或者卡纸上。

③ Pencil/ 铅笔。

Use hard pencil (2H) for drafting patterns, colored pencils for outlining complicated areas.

使用硬质铅笔（2H）绘制纸样，使用彩色铅笔勾画复杂的区域。

④ Fibre pen / 纤维笔。

It is used for writing clear instructions on paper patterns.

用于在纸样上做出清晰的标注。

⑤ Rubber / 橡皮。

⑥ Metric ruler / 公制直尺。

Metric Ruler is used for drawing straight lines and measuring short distances. Generally, it is 20 cm, 50 cm and so on.

绘制直线及测量较短直线距离的尺子，其长度有 20cm、50cm 等数种。

⑦ Curve rules / 曲线板。

It is used for drawing long curves.

用于绘制长的曲线。

⑧ Set square / 三角尺。

A large set square with a 45° angle is very useful. metric grading squares can be obtained.

大的 45° 三角尺是必需的。另外还可以准备一个公制的放码三角尺。

⑨ 1:4 or 1:5 scale square ruler / 1:4 或 1:5 比例尺。

These are essential for students to record pattern blocks and adaptations in their notebooks.

用于在笔记本上记录纸样和进行修改。

⑩ Metric tape measure / 公制皮尺。

⑪ *Metric square ruler / 公制直角尺。

⑫ *French curve ruler / 法式弯尺。

Plastic shapes and curves are available in a range of sizes, and they are useful for drawing good curves.

塑料形状和曲线有各种尺寸可供选择，它们有助于绘制良好的曲线形态。

⑬ Compasses / 圆规。

Compasses are used for drawing circles.

画圆用的绘图工具。

⑭ Tracing wheel / 点线轮。

⑮ Shears / 剪刀。

Different shears should be used for cloth cutting and paper cutting as cutting paper will blunt the blade of cloth cutting.

裁剪布料和纸样应该使用不同的剪刀，因为使用裁剪布料的剪刀裁剪纸样容易弄钝刀片。

⑯ Sellotape / 透明胶带。

⑰ Pins / 大头针。

⑱ Stanley knife / 裁纸刀。

⑲ Tailors chalk / 画粉。

For marking out the final pattern onto the cloth and for marking fitting alterations.

用于在布料上画出裁剪样板和标记适当的更改。

⑳ Toile fabrics / 样衣面料。

Calico is used for making toiles for designs in woven fabrics. Make sure that the weight of the calico is as close to the weight of the cloth as possible. The knitted fabric of the same stretch quality must be used for making toiles for designs in jersey fabrics.

白坯布用于制作梭织面料的样衣。确保白坯布的克重尽量与成衣布料的克重相近。针织面料设计采用与之弹性相同的针织物制作样衣。

㉑ *Model stand / 人台。

Although not essential for a beginner, it is invaluable to a serious student for developing design learning.

对初学者不是必需的，但是对进一步学习设计非常有帮助。

㉒ *Pattern notcher / 牙口剪。

This is a tool that marks balance points by snipping out a section of pattern paper.

用于标记对位点的工具，在纸样上剪下一小块作为对位点。

㉓ *Pattern punch / 打孔器。

㉔ *Pattern weights / 纸样压铁。

It can keep pieces of patterns in position on paper or cloth.

用于把纸样固定在纸或者布料上。

㉕ *Computer equipment / 计算机。

㉖ *Calculator / 计算器。

A calculator is now a common tool in all fields of skill.

计算器是一种适用于所有领域的常用工具。

三、Marks for Making Patterns / 制图符号

The common marks for making patterns are shown in Table 1-4-1.

常用服装结构制图符号如表 1-4-1 所示。

Table 1-4-1 Marks for Making Patterns
表 1-4-1 制图符号

Number （序号）	Sign （符号形式）	Name （名称）	Instruction （说明）
1	▬▬▬▬	Heavy Line （粗实线）	The outlines of the patterns （样板轮廓线）
2	────	Thin line （细实线）	The basic lines of the patterns, the dimension lines and the extension lines （样板结构的基本线，尺寸线和尺寸界线，引出线）
3	- - - - - - - -	Dotted Line （虚线）	The outlines of the components that be covered, sometimes being used to represent the axis of symmetry （被遮住的零部件轮廓线，有时也用来表示对称轴）
4	─ ·· ─ ·· ─ ·· ─	Chain Line （点画线）	Fold line （对折线）
5	△ 2	Special Seam Allowance （特殊放缝）	Different from common seam allowance （与一般缝份不同的缝份量）
6		Zipper （拉链）	Marking the position of the zipper （用于标记装拉链的部位）
7		Radial Mark （经向）	Straight lines with arrows indicate the warp direction of the fabric （带箭头直线表示布料的经纱方向）
8		Direct Mark （顺向）	A mark indicating plush direction on the surface of clothing material. The direction of the arrow is the same as that of the plush. （表示服装材料表面毛绒顺向的标记。箭头方向与毛绒倒向一致）
9		Bias （斜料）	The straight line with arrows indicates the direction of the warp yarns of the fabric （用有箭头的直线表示布料的经纱方向）
10		Dart （省道）	Sew in the dart （将某部位收省）
11		Dart Position Mark （开省符号）	Trim the dart （需剪开省道的部位）

12		Inverted Pleat （阴裥）	Pleat backing in the bottom （裥底在下的褶裥）
13		Visible Pleat （明裥）	Pleat backing in the top （裥底在上的褶裥）
14		Single Pleat （单向褶裥）	Fold up the pleats along the bias （表示顺向褶裥自高向低的折倒方向）
15		Double Pleat （对合褶裥）	Fold up the pleats along the bias （表示对合褶裥自高向低的折倒方向）
16	○ △ □	Equal amount （等量号）	Having the equal amount （两者相等量）
17		Equation Line （等分线）	Divide the line segments in equal proportions （将线段等比例划分）
18		Vertical Mark （直角）	Both in a vertical state （两者成垂直状态）
19		Overlapping Mark （重叠）	Both are overlapping （两者相互重叠）
20		Shrink by sewing （缩缝）	Shrink one piece of fabric when sewing （用于布料缝合时收缩）
21		Shrink Mark （收缩／归拢符号）	Shrink one piece of fabric when sewing or ironing （将某部位收缩／归拢变形）
22		Stretch Mark （拉伸／拔开符号）	Stretch one piece of fabric when sewing or ironing （将某部位拉展变形）
23		Button Position Mark （纽眼符号）	Mark the positions of buttons （两短线间距离表示纽眼大小）
24		Buttonhole Position Mark （扣位符号）	Mark the positions of buttonholes （表示钉扣的位置）

25	⊗ ◎	Press-Button （按扣）	Mark the position of press-button （表示按扣的位置）
26	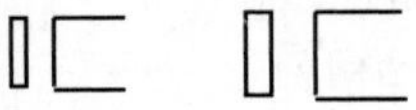	Hook （钩扣）	Set two pieces by hooks （两者成钩合固定）
27	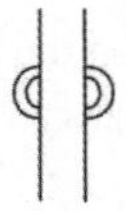	Piece Together Mark （拼合符号）	Piece together related lines （表示相关线条拼合一致）
28	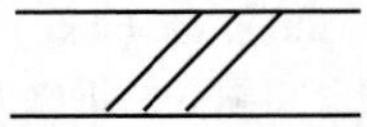	Interlining （衬布）	Interlining （表示衬布）
29	（前） （后）	Balance Mark （对位记号）	These are used to make sure that pattern pieces are sewn together at the correct points （用来保证纸样各裁片能够正确缝合在一起的标记点）
30	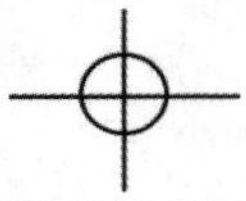	Drilling Mark （钻眼位置）	Mark the position of drilling when cutting （表示裁剪时需钻眼的位置）

Note: If other special or non-standard marks are used when making patterns, they must be instructed with diagrams and text in the drawings.

注： 在制图中，若使用其他制图符号或非标准符号，必须在图纸中用图和文字加以说明。

四、Codes for Patterns / 服装制图中的主要部位代号

Codes for Patterns are shown in table 1-4-2.

服装制图过程中使用的主要部位代号如表 1-4-2 所示。

Table 1-4-2 Code for Patterns
表 1-4-2 服装制图中的主要部位代号

Number （序号）	Critical points and lines in patterns （纸样上的关键点和线）	Code （代号）
1	Bust Point （胸点）	BP
2	Side Neck Point （侧颈点）	SNP
3	Front Neck Point （前颈点）	FNP
4	Back Neck Point （后颈点（第七颈椎点））	BNP
5	Shoulder Point （肩端点）	SP
6	Front Center Line （前中心线）	FCL
7	Back Center Line （后中心线）	BCL
8	Neck Line （领围线）	NL
9	Chest Line （上胸围线）	CL
10	Bust Line （胸围线）	BL

11	Under Bust Line（下胸围线）	UBL
12	Chest Width Line（胸宽线）	CWL
13	Back Width Line（背宽线）	BWL
14	Waist Line（腰围线）	WL
15	Middle Hip Line（中臀围线）	MHL
16	Hip Line（臀围线）	HL
17	Elbow Line（肘线）	EL
18	Sleeve Opening Line（袖口线）	SOL
19	Knee Line（膝盖线）	KL
20	Head Size（头围）	HS
21	Neck Girth（领围）	N
22	Front Neck（前领围）	FN
23	Back Neck（后领围）	BN
24	Bust Girth（胸围）	B
25	Waist Girth（腰围）	W
26	Hip Girth（臀围）	H
27	Thigh Size（大腿根围）	TS
28	Length（衣长）	L
29	Front Length（前衣长）	FL
30	Back Length（后衣长）	BL
31	Front Waist Length（前腰节长）	FWL
32	Back Waist Length（后腰节长）	BWL
33	Front Chest Width（前胸宽）	FCW
34	Back Width（后背宽）	BW
35	Trousers Length（裤长）	TL
36	Skirt Length（裙长）	SL
37	Inside Length（股下长）	IL
38	Body Rise（股上长 / 立裆）	BR
39	Front Rise（前裆弧长）	FR
40	Back Rise（后裆弧长）	BR
41	Slack Bottom（脚口）	SB
42	Armhole（袖窿）	AH
43	Armhole Depth（袖窿深）	AD
44	Arm Top（袖山顶点）	AT
45	Biceps Circumference（上臂围）	BC
46	Cuff Width（袖口宽）	CW
47	Sleeve Length（袖长）	SL
48	Elbow Length（肘长）	EL
49	Collar Band（领座）	CB

50	Collar Rib（领高）	CR
51	Collar Length（领长）	CL

五、Names of Lines for Patterns / 纸样上各部位线条名称

1. Names of lines for top patterns / 衣身上的线条名称

(1) Basic lines for top patterns / 衣身基础线

Basic lines for top patterns are shown in Fig. 1-4-1.

衣身基础线如图 1-4-1 所示。

(2) Construction lines for top patterns / 衣身结构线

Construction lines for top patterns are shown in Fig. 1-4-1

前后衣身结构线如图 1-4-1 所示。

1—Top Flat Line / 上平线

2—Top Length Line / 衣长线

3—Front Center Line / 前中心线

4—Front Edge Line / 门襟止口线

5—Back Center Line / 后中心线

6—Front Neck Width Line / 前领宽线

7—Front Neck Depth Line / 前领深线

8—Back Neck Width Line / 后领宽线

9—Back Neck Depth Line / 后领深线

10—Shoulder Slop Line / 落肩线

11—Shoulder Line / 肩宽线

12—Chest Width Line / 前胸宽线

13—Back Width Line / 后背宽线

14—Bust Line / 胸围线

15—Waist Line / 腰围线

16—Hip Line / 臀围线

17—Side Seam Line / 侧缝线

18—Front Finish Line / 撇门线

19—Line of Front Cut Point / 门襟圆角线

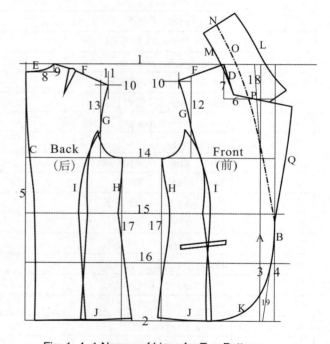

Fig. 1-4-1 Names of Lines for Top Patterns

图 1-4-1 衣身上的线条名称

A—Front Center Line / 前中心线

B—Front Edge Curve / 门襟止口弧线

C—Center Back Seam Line / 背缝线

D—Front Neck Line / 前领窝线

E—Back Neck Line / 后领窝线

F— Shoulder Line / 肩斜线

G—Armhole Line / 袖窿线

H— Side Seam Line / 侧缝线

I—Split Line / 分割线

J—Hem Line / 下摆线

K— Line of Front Cut Point / 门襟圆角线

L—Outside Line of Collar / 领子外口线

M—Under Line of Collar Band / 领座下口线

N— Back Midline of Collar / 后领中线

O—Fold Line of Collar / 领子翻折线

P—Gorge Line / 串口线

Q—Lapel Edge Line / 驳头止口线

2. Names of lines for sleeve patterns / 衣袖上的线条名称

(1) Basic lines for sleeve pattern / 衣袖基础线

Basic lines for sleeve patterns are shown in Fig. 1-4-2.

衣袖基础线如图 1-4-2 所示 .

(2) Construction lines for sleeve patterns / 衣袖结构线

Construction lines for sleeve patterns are shown in Fig. 1-4-2.

衣袖结构线如图 1-4-2 所示 .

1—Top Flat Line / 上平线

2—Sleeve Length Line / 袖长线

3—Sleeve Crown Line / 袖山线

4—Elbow Line / 袖肘线

5—Up Line of Sleeve Opening / 袖口翘线

6—Sleeve Crown Length Line / 袖山高线

7—Front Sleeve Seam Line / 前袖缝直线

8—Fold Line of Top Sleeve / 前偏袖线

9—Fold Line of Under Sleeve / 后偏袖线

10—Back Sleeve Seam Line / 后袖缝直线

11—Sleeve Center Line / 袖中线

A—Sleeve Opening Line / 袖口线

B—Sleeve Crown Curve / 袖山弧线

C—Front Sleeve Seam Curve / 前袖缝弧线

D—Fold Line of Top Sleeve / 前偏袖线

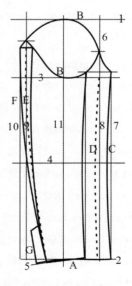

Fig. 1-4-2 Names of lines for sleeve pattern
图 1-4-2 衣袖上的线条名称

E—Fold Line of Under Sleeve / 后偏袖线

F—Back Sleeve Seam Curve / 后袖缝弧线

G—Sleeve Vent Line / 袖衩线

3. Names of lines for bottom patterns / 下装样板上的线条名称

(1) Basic lines for bottom patterns / 下装基础线

Basic lines for bottom patterns are shown in Fig. 1-4-3.

下装基础线如图 1-4-3 所示。

(2) Construction lines for bottom patterns / 下装结构线

Construction lines for bottom patterns are shown in Fig. 1-4-3.

下装结构线如图 1-4-3 所示。

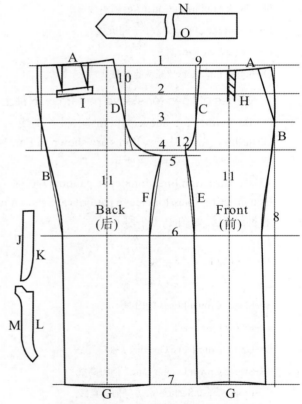

1—Top Flat Line / 上平线

2—Middle Hip Line / 中臀线

3—Hip Line / 臀围线

4—Crutch Depth Line / 横裆线或立裆深线

5—Downcast Line / 落裆线

6—Knee Line / 中裆线

7—Trousers Length Line / 裤长线

8—Side Seam Line / 侧缝线

9—Line of Front Rise / 前上裆斜线

10—Line of Back Rise / 后上裆斜线

11—Trousers Center Line / 挺缝线

12—Front Rise Width Line / 前裆宽线

A—Waist Seam Line / 腰缝线（腰口线）

B—Side Seam Line / 侧缝线

C—Curve Line of Front Rise / 前上裆弧线

D—Curve Line of Back Rise / 后上裆弧线

E—Front Inside Seam Line / 前下裆线

F—Back Inside Seam Line / 后下裆线

G—Bottom Line / 脚口线

H—Pleat Position Line / 褶位线

I—Pocket Position Line / 袋位线

J—Inside Line of Left Fly / 裤门襟止口线

K—Outside Line of Left Fly / 裤门襟外口线

L—Inside Line of Right Fly / 裤里襟止口线

M—Outside Line of Right Fly / 裤里襟外口线

Fig. 1-4-3 Names of lines for bottom patterns
图 1-4-3 下装上的线条名称

N—Top Line of Waistband / 腰头上口线

O—Finish Line of Waistband / 腰头下口线

Unit 5 Pattern making tools in CAD Software
第五节 CAD 软件中的制板工具

This book uses *Zhizunbaofang* CAD Software, it is suitable for any style of structural drawings. There are 10 ways of point-capture alone, nearly 30 tools of point and line, and more than 100 functions in total. It displays the power of the computer—freely drawing, modifying curves, editting point, automatic smoothing. It is convenient, quick, intuitive to use. It has complete industrial symbols. The window is shown in Fig. 1-5-1.

本书采用《智尊宝纺服装CAD》软件，它适用任何造型的结构制图。单独的点捕捉方式就有 10 种，点、线工具近 30 种，加上其他的工具，总共有 100 多种功能可供使用。它展示了计算机的力量——自由绘制、修改曲线、编辑点、自动圆顺。它操作方便、快捷、直观，拥有齐全的工业符号。其操作界面如图 1-5-1。

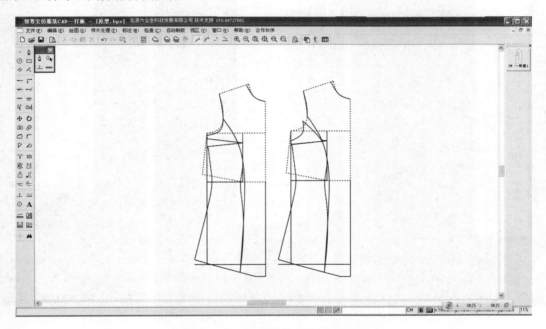

Fig. 1-5-1 Operation interface of CAD
图 1-5-1 软件操作界面

一、Menu / 菜单

(1) There are the following common tools in the menu of "Document".
在"文件"菜单中主要包含下列常用工具。

① New: Produce a new work area. Double-click the mouse to select or delete the size and names of parts of clothes in the dialog box – select "basic size and grade" in the specifications table – click "divide all parts" – click "divide" – OK.

新建：新建一个工作区。在对话框中通过双击鼠标选择或删除尺码及部位名称—在规格表中选择"基准码尺寸和档差"—点击"等分所有部位"—点击"等分"—确定。

② Open: Open an existent pattern.

打开：打开一个已经存在的样片文件。

③ Save: Save the pattern in the work area. Every pattern includes three files, having the same name but different suffix: *.bps, *.pis, *.psn.

保存：保存工作区内的样片文件。每个样片包含三个同名不同后缀的文件： *.bps、*.pis、*.psn。

④ Save as: Save an old document as another name.

另存为：将一个已经保存过的文件用其他名称保存。

(2) Other tools in the menu are shown in Fig. 1-5-2.

其他菜单所包含的工具如图 1-5-2 所示。

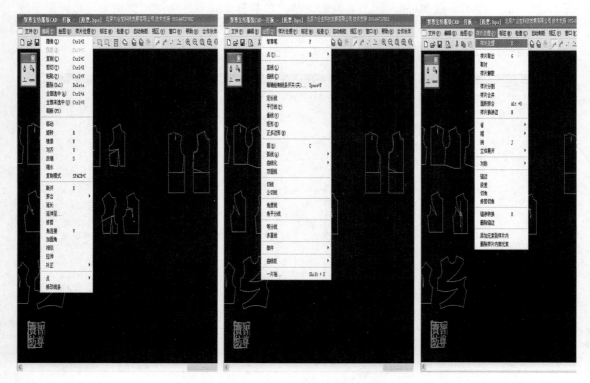

Fig. 1-5-2 Tools in Menu

图 1-5-2 菜单所包含的工具

① Edit: Operate the selection, including Copy, Move, Rotation, Symmetric, etc.

编辑：对选定的对象进行操作，包括复制、移动、旋转、对称等。

② Draw: It includes tools of drafting. It is used to draw various points and lines.

绘图：内含制图工具，用来画各种点、线。

③ Handle pattern: Operate the existing pattern, including Split, Merge, Seam Allowance, etc.

样片处理：对已经产生的样片进行操作，包括样片分割、合并、加缝份等。

④ Label: Make labels on the pattern, such as Notch, Grain Line, Words, etc.

标注：对样片进行标注，包括刀口、纱向、文字等。

⑤ View: Set the interface or type of lines in system settings. In this CAD Software, there are some ways to select objects such as click selection, frame selection, reverse-frame selection, and all selection. Right-click on the selected object may cancel the selection. There are two tools in the shortcut toolbar. Which are "select all (▢)" and "cancel to all select (▢)".

视区：对本系统界面的设置，也可在"系统设置"中对线段的属性进行设置。本软件中，选择操作对象的方式有：点选、正框选、反框选、全选几种，取消选择时可以用右键在选中的对象上单击。在快捷工具栏中有全选、取消选择两个工具： ▢ 、 ▢ ，鼠标左击就可以执行相应功能。

⑥ Open "View – System set – Option": Set basic information of the system, parameter input or output devices, display or hide some box of tools etc. The user can set specific requirements.

打开"视区—系统设置—选项"：可以设置系统基本信息、输入和输出设备属性、各类工具栏的显示与隐藏等。用户可根据需要进行设置。

二、Operation of an empty state and common tools / 空状态及常用工具的操作

Common tools box consist of four necessary tools, which are "Zhizun pen", "Handle pattern", "Label" and "Measurement". They are found in the menu or draw tool-box too. You can select these tools easily.

常用工具主要由操作频率比较高的四个工具组成，分别是"智尊笔""样片处理""标注""测量"，在菜单栏或绘图工具栏中也能找到相应的工具，可随意选用。

1. Empty state / 空状态

An empty state is a state of no tools being selected. In the empty state, some ways of operating are as follows:

空状态是指没有打开任何工具时的状态。空状态下可进行以下一些操作：

(1) Modify curves, move a dot / 修改线形状、移动点

Double-click a line and click a dot on the line – move the mouse to a suitable position – click it. If selecting the endpoint, you can move one or more dots. Press the "Tab", the curve is fine-tuned by adjusting the curvature.

双击线段，单击线段上的点，移动鼠标到适合的位置，再次单击。如果选择端点则可以移动一个点或多个点。按 Tab 键，可以通过调节切矢微调弧线。

(2) Edit line / 编辑线段

Select a line – double-click in the blank space (Double-click is the default meaning to double-click the left button) – press the "Tab", you can move or mirror or alignment or rotate the line according to the prompt in the status bar.

选择线段—在空白处双击（双击是默认指双击左键）—按 Tab 键切换，可按状态栏的提示对线段进行移动、镜像、旋转、对齐等操作。

(3) Move pattern / 移动样片

Move the pattern to the suitable position with dragging the center of pattern by using mouse.

点住样片中心拖动，可将样片移动到适合的位置。

(4) Move the grain line of pattern / 移动样片丝缕

Double-click the center of the grain line of pattern, you can move the grain line of pattern to the suitable position and click to ensure.

双击样片丝缕纱向中心，可以移动丝缕标记到满意的位置，单击定位。

(5) Shift among other tools / 其他工具之间的切换

In an empty state, double-click – "Zhizun pen" appears, then double-click – "Rectangle" appears and right-click – "Zhizun pen" appears, and then right-click – in an empty state.

空状态下双击—变智尊笔，双击—变矩形，单击右键—变智尊笔，再单击右键—变空状态。

There are four ways of shifting between tools : double-click, right-click, press the "Tab" and intelligent imagine.

各大类工具之间的切换都有：双击、右击、按"Tab"键、智能联想四种方式。

2. Zhizun pen / 智尊笔 ✒

It is mainly used for drawing and editing straight lines or curves in structural drawings. It is a different function because of the different positions on a line. (The drawing steps in this book default to those in Zhizun Pen mode).

主要用于结构图中直线或弧线的绘制及编辑。当智尊笔位于线上时，因所处的位置不同，作用也不同。（本书中的制图步骤默认为智尊笔模式下的步骤）。

(1) Draw a line / 绘制线段

Click (left-click by default) each point consecutively with mouse or click each point captured by input value to form a curve and right-click to switch the curve to a straight line. It is over when right-click again. The horizontal or vertical lines can be drawn when we press "Shift" in the meantime.

用鼠标连续点击（默认为击左键）各点或点击输入数值捕捉到的各点连成弧线，点击右键可切换弧线成直线，再按右键可结束绘制线段。按住 Shift 不放时可作水平线、垂线。

There are four ways to add points or capture points on a line based on the data: fixed length ⟋ , ratio ⟋ , relative ⤻ , projector ⊥ . Select tool, input data according to the prompt in the status bar.

根据数据在线段上加点或捕捉点的方法有四种，即定长点（ ⟋ ）、比例点（ ⟋ ）、相对点（ ⤻ ）、投影点（ ⊥ ），选择相应的工具，按状态栏提示输入数据即可。

(2) Lengthening or shortening a line / 延长或缩短线段 ⊢⋯

Select a line – move the mouse to the end of the line, click (left-click by default) – input data – enter. (When the input number is positive, the line is lengthened, and when it is negative, the line is shortened).

选择一条线段，将鼠标移动到线段端点—单击（默认为击左键）—数据栏输入延长或缩短的数据—回车。（当输入数为正值则线段延长，输入数为负值则线段缩短）。

(3) Parallel line / 平行线 ∥

Select a line – move the mouse to the line and "parallel" appears – click – input data – enter.

选择一条线段，将鼠标移动到线段上出现平行符号—单击—输入数据—回车。

(4) Break off a line / 断开线段 ⊷

Select a line – move the mouse to the point of intersection or input data, and a knife symbol appears – click.

选择一条线段—将鼠标移动到线段相交点处或输入数值出现刻刀符号—单击。

(5) Vertical line / 垂直线 ⟨入⟩

Select a line – move the mouse to the line – right-click, and a vertical symbol appears – select the vertical point position or enter data in the data bar – select another point or enter data.

选择一条线段—将鼠标移动到线段上—右击出现垂直符号—选择垂点位置或在数据栏输入数据—选择已知点或输入数据。

(6) Dig a dart, break a dart, transfer a dart / 挖省、掰省、转省 ⟨V⟩

Dig a dart: Select a line – move the mouse to the intersection line – a dart symbol appears – click it – input data – enter.

挖省：选择一条线段，将鼠标移动到相交线上出现省道符号—单击相交线—输入数据—回车。

Break a dart: Select a line to make dart – select its centerline and dart tip – select the line that will be moved (near the fixed end) – right-click – input dart width – enter.

掰省：选择开省线—选择省中心线、省尖点—选择参与移动的线段（靠近线段不动的一端）—右击—输入省宽—回车。

Transfer a dart: e.g. transfer side seam dart to split line. Frame select two dart lines – move the mouse to the split line, click – select the segment between the split line and dart – right-click.

转移省：例如将侧缝省转移到分割线。框选二条省边线—移动鼠标到分割线，单击—选择分割线与省道之间的线段—右击。

(7) Create a similar line / 做相似线 ⟨⊘⟩

Select a line – input one end width – click its adjacent line – enter another end width – click another adjacent line.

选择一条线段—输入一端宽度—单击对应邻线—输入另一端宽度—单击对应邻线。

(8) Lengthening to a line / 延长至已知线段 ⟨–⟩

Select a line – move the mouse to the line to be lengthened – click.

选择已知线段—将鼠标移动到需要延长的线段—单击。

(9) Trim a line / 修剪线段 ⟨–⟩

Select intersection lines – move the mouse to the line to be trimmed, click.

选择二条相交线段—将鼠标移动到需要剪去的部分线段，单击左键。

(10) Splice, correction / 拼合修正 ⟨⊫⟩

Select two adjacent lines – move the mouse to the line to be corrected, right-click.

选择二条相邻线段—将鼠标移动到需要修正的部分线段，单击右键。

(11) Angle bisector / 角平分线 ⟨∠⟩

Select two intersection lines – move the mouse to the point of intersection, click.

选择二条相交线—将鼠标移动到交点处—左击，即可作角平分线。

(12) Make a fillet / 加圆角 ⟨⌐⟩

Select intersection lines – move the mouse to the point of intersection, right-click.

选择相交的线—将鼠标移动到交点处，右击。

3. Handle and edit pattern / 样片处理与编辑

Handle pattern and make or edit dart, pleat, tuck on patterns.

用于对样片的处理以及样片上省、褶、裥的生成和编辑。

1）Handle pattern / 样片的处理

(1) Take out pattern / 取出样片

Double-click the blank space – select a closed outline – right-click – set the corresponding content in the dialog box – Determine.

双击空白处—依次选择封闭的轮廓线—右击—在对话框中设置相应内容—确定。

(2) Create a symmetric pattern / 产生对称样片

Double-click the symmetry axis on the pattern.

在样片上双击对称轴即可。

(3) Split pattern / 分割样片

Select a pattern – move the mouse to the split line and "scissors" appears – click all split lines – right-click.

选中样片—靠近分割线，出现剪刀工具—单击所有分割线—右击。

(4) Merge two patterns together / 样片合并

Select a merge line on one pattern – right-click – select a merge line on another pattern – right-click.

选择一个样片的合并线—右击—再选择另一个样片的合并线—右击。

(5) Add seam / 加缝边

Select lines or a pattern – right-click – input seam allowance – enter.

选择线段或样片—右击—输入缝份量—回车。

(6) Cutting angle / 切角

Select net lines (several line segments can be selected at once) – click again when "cutting angle" appears on the line – select the type of angle – click "apply".

选择净边线段—光标在线上出现切角工具时再次单击—选择切角类型—点击"应用"。

(7) Trim cutting angle / 修剪切角

It applies to cutting corresponding seam corners of two pattern pieces. Select one net line – select the other net line – move the mouse on the line, click when "automatic trim " appears – click.

适用于两个样片的对应缝份角的修剪。选择一净边—选择另一净边—光标放在线上变自动修角工具时单击—左击确定。

(8) Segment difference / 段差

Select one net line of pattern – move the mouse to its end and click when "vent " appears – set the type and parameters – Determine.

选择一个样片的净边线—移动光标至该净边线端点时，出现"段差"时单击—选择类型及输入参数—确定。

2）Handle dart on pattern / 样片上省的处理

First, dig a dart on the pattern with Zhizun pen. Then right-click the dart line in an empty state and a dart menu appears. Use it to edit, turn, smooth, delete and add a dart hill to a dart.

首先用"智尊笔"在样片上挖省，然后在空状态下右击省边线将出现省道菜单，利用该菜单可进行省编辑、省转移、省圆顺、加省山、删除省等操作。

(1) Edit dart / 省编辑

Used to set dart parameters. Select a dart line – a dialog box appears – input parameters – Determine.

用于设置省参数。选择省边线—出现省修改对话框—输入参数编辑完成后—确定。

(2) Ratio transfer dart / 比率转省

Used to transfer a dart proportionally to other position and the dart tip hardly moves. Select a dart tip – select a dart line – select a new dart line anti-clockwise in turn, right-click – enter every ratio – enter.

用于将省按比率地转移到其他位置，省尖位置基本不变。选择省尖—选择省线—按逆时针方向依次选择新省线，右击结束—输入各省比率值—回车。

(3) Equal transfer dart / 等分转省

Used to transfer a dart equally. The new dart tip must be on the old dart line and its opening must be on the outline. Select the old dart – select the new dart line, right-click – enter the type of smoothing – enter.

用于将省等分地转移。新省尖位置必须在旧省边线上，其开口在样片轮廓上。选择待转省—选择新省线，右击结束—输入圆顺种类—回车。

(4) Merge dart / 合并省

Used to merge one dart to another and shut its opening. Select a dart to be shut – select the other dart – select the center. (It's only one dart being merged every time.)

将一个省道合并到另一个省中，原来位置闭合。选择待闭合省—选择并入的省—选择省中心。（每次只能合并一个）

(5) Smooth line being opened dart / 省圆顺

Used to modify lines that being build up a dart. It can smooth lines. Select the dart which will be smoothed—right-click – select the fixed dart line – modify (move along the dart line with pressing "Ctrl". The dart will be modified with the curve without pressing "Ctrl".) – left-click to to confirm, right- click to cancel.

用于开省后的线段修改，使其圆顺。选择需要圆顺的省—右击—选择不动侧—修改闭合处曲线造型（按 Ctrl 键，省沿省位线移动；不按 Ctrl 键，省会随着曲线的改变而改变）—左击确定，右击取消。

(6) Curve dart / 省道曲线化

Turn darts from straight to curved. Select a dart – move the mouse to do that operation – left-click to confirm.

将直线省转化为曲线省。选择省—移动鼠标—左击定位。

(7) Definition dart / 定义省

Two intersection lines on the pattern are defined as dart lines. Only the dart that has been defined can be edited or transfered. Select the first line anti-clockwise – right-click – select the second line –

right-click to end.

将样片上二条相交线段定义为省道的二条边线，定义后的省才可以进行转省、省编辑等操作。逆时针方向选择第一条线—右击—选择第二条线—右击结束。

3）Handle pleat of pattern / 样片上褶、裥处理

Making Pleat: In "handle pattern" state, select the center of the pattern (default is the center of the grain line on the pattern) – move the mouse to the auxiliary line and "scissor" appears – click – press "Tab" twice and "pleat" appears – select other lines near the endpoint – right-click to end – a dialog box of "pleat" appears – set the parameters – Determine – select "pleat towards".

褶的生成："样片处理"工具下，选择样片中心（默认为样片的纱向中心）—移动鼠标到辅助线上，出现"剪刀"图标—单击—按 Tab 二次切换到"褶裥"图标—在近端点处选择剩余褶线—右击结束—出现"褶"对话框—选择相应参数—确定—选择"褶的倒向"。

Edit Pleat: The pleat can be modified. In empty state, select the pattern – move the mouse to the symbol of pleat – right-click – pleat options such as Edit, Open, Shut, Delete and Dissolve appear. You can choose one to operate.

褶的编辑：褶生成后，可以进行修改。空状态时选择样片 — 移动鼠标到褶标注符号上 — 右击 — 出现"褶编辑、褶展开、褶闭合、褶删除、褶解散"等选项，可分别选择进行相应操作。

If you choose "pleat", you can turn a pleat into a short pleat. Select the end being turned – enter its length.

选择选项"裥"，则可以将褶转化为短裥。选择转化为裥的一端—输入裥长。

4）Open pattern / 样片展开

(1) Cut and move pattern / 样片剪开与移动

In "handle pattern" state, select the center of the pattern and move the mouse to the auxiliary line and "scissors" appears – click – press "Tab" once and "cut and move pattern" appears – near the cut end, click the cut line in turn, and right-click to end – set parameters – select base points – Determine.

"样片处理"工具下，选择样片中心，移动鼠标到剪开线上出现"剪刀"图标，单击—按 Tab 键切换到"剪开 / 移动"工具—偏向剪开的一端依次单击剪开线，右击—输入相关参数—选择基准点—确定。

(2) Equal cut and move pattern / 等分割移动

In "handle pattern" state, select the center of the pattern – select the first group lines, right-click – select the second group lines, right-click – a dialog box appears – set parameters – Determine.

"样片处理"工具下，选择样片中心—选择分割的第一组线段，右击—选择分割的第二组线段，右击—出现对话框，设置相应数值—确定。

(3) Open both sides / 两侧展开

For example, make puff sleeve. In "handle pattern" state, select the base lines – select the opening lines, right-click – input data in the dialog box – enter "view" – Determine.

如制作泡泡袖。"样片处理"工具下，选择基础线段—选择展开线，右击—对话框内输入数值，回车可预览—确定。

Smooth line and replace. In "handle pattern" state, select the old lines, right-click – select the new lines, right-click – then delete the inner elements of the pattern.

修顺线并替换。"样片处理"工具下，选择原来线，右击结束—选择新线，右击结束，即替换了原来的线—再删除样片的内部元素。

(4) Open one side / 单侧展开

Similar as open both side. In "handle pattern" state, select base lines – select fixed lines, the mouse turn into "open one side" – select opening lines ,right-click – input data in the dialog box – enter view – Determine.

与"两侧展开"类似，进行单边展开。"样片处理"工具下，选择基础线段—选择固定边，鼠标变为单侧展开状态—选择展开线，右键结束—对话框内输入数值，回车可预览—确定。

(5) Replacement net outline / 样片换净边

In "handle pattern" state, select the unsuitable lines – right-click – select the new lines – right-click.

"样片处理"工具下，选择需要替换掉的边线，右击结束—选择新线，右击结束，即可替换掉不需要的线。

4. Label / 标注

Make or edit labels on the pattern.

用于样片上制作或编辑标记。

(1) Modify grain line / 修改纱向

Select the pattern – click start and end point. Press "Shift" and you can draw a horizontal or vertical grain line. Press "Tab" and you can draw a 45º direction grain line. If you want to draw a grain line parallel to a line, you can select the pattern and select the line – draw the line.

选择样片—单击起点和终点即可作任意方向丝缕线。同时按 Shift 键，可以作水平或垂直方向的纱向，按 Tab 键可作 45º 方向丝缕。如果需要作与某一线段平行的纱向时，可选择样片，再选择要平行的线—画纱向即可。

(2) Set button position / 定纽位

Select the line – move the mouse to the auxiliary line and "button" appears – click, the dialog box of button appears – input data – Determine – right-click. (Left-click, and buttonhole will reverse.)

选择线段—移动鼠标到线上，出现纽扣符号—左击，出现纽扣属性对话框—输入相关数据—确定—右击结束。（扣眼反转按左键）

(3) Label pressing line / 标压线

Select the line – move the mouse to the auxiliary line and "button" appears – right-click, the dialog box of pressing line appears – input data – Determine.

选择线段—移动鼠标到线上，出现纽扣符号—右击，出现压线属性对话框—输入相关数据—确定。

(4) Label gathered / 标波浪褶

Select the line – right-click in the blank space – select start and end point – right-click.

选择线段—空白处右击—选择起、止点位置—右键结束。

(5) Add words / 添加文字

Double-click in the blank space – an "A" will appear – make a frame and a dialog box will appear – input words or set their sizes – Determine. To add words or elements to the pattern, in empty state, right-click the center of the grain line on the pattern – select "add element to pattern" – select the content – right-click.

双击空白处—出现文字图标—拉框出现文字对话框—输入文字、设置字体大小—确定。如果要在样片内添加文字或元素，则先在空状态下右击样片上纱向中心处—"添加元素到样片内"—选择要添加的内容—右击。

(6) Insert or edit pitch point / 刀口插入及编辑

Insert Pitch Point　Select a line and "pitch point" will appear – click or enter data, enter. Right-click and a dialog box of pitch point will appear – input data – Determine.

靠近线段—出现刀口图标—单击或输入数据回车。如果在出现刀口图标时右击，则出现刀口"属性设置"对话框，输入相关数据后—确定。

Edit Pitch Point　Click "pitch point" – create an empty state – right-click and "edit option of pitch point" appears – choose an option.

在刀口位置单击，光标切换到空状态，这时右击会出现"刀口编辑选项"，选择选项即可进行编辑。

(7) Coloring pattern / 样片填色

Select the pattern – right-click the center of the pattern – a dialog box appears – double-click "style" – Determine.

选择样片—右击样片中心—出现对话框—双击"样式"选择—确定。

(8) Delete label / 删除标注

Select the center of the pattern in empty state – right-click – choose "delete label" – choose the label. If you want to delete all labels, select the label line press Ctrl – left-click; If you want to delete a single label, select the label line – left-click, can delete pressure line, gathers line, buttons position, etc.

空状态下选择样片中心—右击—选择"删除标记"—选择要删除的标记。全部删除则按住 Ctrl 选择做标记的线—左击；单个删除则选择线段—左击，可删除压线、波浪褶、扣位等。

5. Measurement and check / 测量检查

(1) Distance measurement / 测点距（直线）

Used to measure the total length between two points or some points. Click start point in turn, right-click to end.

用于测量二个点或多个点之间线段总长度。依次单击各线段的起始点，右击结束。

(2) Curve measurement / 测弧线长

Select a line – select two points on the line. If measuring a line or several lines, you can select the line or lines – click or click in succession.

选择线段—选择线段上的两点。如果要测量一条或几条线段长，则选择该线段—单击或连续单击。

(3) Angle measurement / 测量角度

Select two lines of an angle in counterclockwise order.

按逆时针顺序先后选择角的两边。

(4) Alignment check / 对位检查

For example, two pitch points are aligned or not on two line segments. Double-click – select the start and end pitch points on the first line segment – right-click – right-click to end; Select the start and end pitch points on the second line segment – right-click – right-click to end – a table of "length and differ" appears – modifying data value can adjust the pitch point position.

例如二条线段上刀口对齐的检查，双击—选择第一条线段起止对位点—右击—右击结束；选择第二条线段的起止对位点—右击—右击结束—出现二条线段对位点之间的长度情况及差值，改变其中的数值可以调整相应对位点的位置。

(5) Merge Check / 拼合检查

It can be used to compare the lengths of two line segments. Box select the first group line segments, right-click – box select the second group line segments, right-click.

可以比较两条线段的长度。框选第一组线段，右击—框选第二组线段，右击。

(6) Joint check / 接角对合检查

Check the effects of the jointed lines. Press "Tab" to switch to this tool – select the lines – right-click to end and the effect appears, and modify the point position or add or delete points. Press "Tab", and you can change the curvature of a curve. Left-click to Determine and right-click to end.

检查多条线段对接后的效果。按 Tab 键切换到此工具—连续选择对接的线段—右击结束，显示线段对接后的效果，并可以对各点位置进行修改，增加或删除点，按 Tab 键可调节切矢。调节完成后按左键确定，右击结束。

三、Commonly quickly key / 常用快捷键

A：display all sizes / 显示全部尺码。

Space +L：display data or cancel it / 显示数据或取消。

Space +H：display seam allowance or cancel / 显示缝份或隐藏。

空状态下 Q：close to a line segment to find the steps of operation / 靠近线段，查找操作步骤。

Exercise / 练习

1. Remember the names of various parts of the human body /
熟记人体各部位名称。

2. Measure body size for more than 2 students, each girth should be measured three times, and all measured data should be recorded /
给 2 个以上同学进行体型测量，要求每个围度部位测量 3 次，记录所有测量的数据。

3. Memorize the names of various structural lines on clothing /
熟记服装上各种结构线名称。

4. Memorize the meanings of various drawing symbols /
熟记各种制图符号的意义。

5. Master the operation methods of various basic tools in CAD software /
熟练掌握 CAD 软件中各种基本工具的操作方法。

Chapter 2 Basic patterns for women
第二章 女装基础纸样

Unit 1 Methods of creating basic patterns
第一节 获得基础纸样的方法

一、A basic Pattern / 基础纸样

A basic pattern is a foundation pattern for garment structure design. The structure of the basic pattern is the simplest form. The basic pattern, which is constructed to reach the biggest coverage, contains information on the most important parts of the body and reflects the essential features of the body for design. A basic pattern is a transitional form of garment structure and not the final pattern used for the real fabric.

基础纸样是服装结构设计的基础图形，是结构最简单，能包含人体最重要部位信息，能从本质上反映人体主要特征，但力求具有最大覆盖面的服装纸样。基础纸样是服装结构构图的过渡形式，而不是用于实际面料的最终纸样。

Garment structure varies while clothing styles change. In fact, everything changes according to certain rules. Designing garment patterns means that you can create a variety of style patterns as long as the "Basic" (i.e. basic pattern) stays constant to cope with the changes. These changes being: Adjusting the ease, rotating, cutting, folding patterns, etc., to construct darts, pleats, drapes and style lines or seams.

随着服装款式千变万化，服装结构也变得多端。事实上，任何事物的变化都是遵循一定规律的。进行服装纸样设计时，只要抓住"基本型"（即基础纸样），以不变应万变，通过对基础纸样的松量调节、旋转、剪切、折叠等变形方法，采用省道、褶裥、分割、连省成缝等结构形式，便可得到各种款式的服装纸样。

二、Features of a basic pattern / 基础纸样的特点

A basic garment pattern design should have the following features:

作为服装结构设计基础的纸样，应该具有以下特征：

(1) Several measurements need to be obtained simply and accurately.

测量部位少，量取尺寸简单，准确率高。

It is necessary to have accurate human body measurements to ensure the garment fit. However, there will be more measuring errors made by the personal measuring technique level and the measuring tools and other factors if there are more measured parts. It is a disadvantage for making patterns accurately and efficiently.

人体部位尺寸的测量是制作合体服装所必需的。但是，如果测量部位越多，由个人测量技术水平、测量工具等因素造成的误差会越大。这显然不利于基础纸样的准确、高效的制作。

(2) The drawing process is simple and the drawing method is easy to remember, and the formula and constant value adjustment are as seldom as possible so that the pattern makers can manage them.

制图过程简单，制图方法简便易记，尽可能地减少繁琐的计算公式和定数调节，便于制图者掌握。

(3) It is a suitable fit for the human body.

适合人体。

(4) It has advantages of the application and development of various clothing styles.

具有各种变化服装款式的应用、发展优势。

三、Types of basic patterns / 基础纸样的分类

According to their different features, basic patterns can be classified in different ways.

根据基础纸样的不同特点，可以从不同的方面进行分类。

(1) Sorted by sex or age: women's basic patterns, men's basic patterns, and children's basic patterns.

按性别或年龄可分为：女装基础纸样、男装基础纸样、童装基础纸样等。

(2) Sorted by body part: the basic upper body pattern, the basic sleeve pattern, the basic skirt pattern and the basic pants pattern.

按人体躯干部位可分为：上衣基础纸样、袖子基础纸样、裙子基础纸样、裤子基础纸样等。

(3) Sorted by country: basic British pattern, basic American pattern, basic Italian pattern, basic Japanese pattern, etc.

按国别可分为：英式基础纸样、美式基础纸样、意式基础纸样、日式基础纸样等。

四、Methods for attaining a basic pattern / 获得基础纸样的方法

A basic pattern is an essential approach and way for fashion designers to grasp design skills and fashion styles. Different designers have different methods to achieve a basic pattern. At present, there are three methods. These are: the flat method, the draping method and the method of combining flat with draping.

基础纸样是服装设计师把握和设计服装造型的基本途径和手段，不同的设计师获取基本纸样的方法不尽相同，当前，基础纸样的获取有平面法、立体法、立体和平面相结合法三种方法。

1. The flat method / 平面法

The flat method is one of the most popular methods for pattern making. It transfers the three-dimensional form of the clothing to a flat form, which means that it transfers three-dimensional body figures into two-dimensional patterns through the garment and then it creates a flat pattern according to certain numbers or formulas. It takes advantage of patternmakers' long-term experience and science development and becomes simple, convenient, economical and practical, scientific and efficient. It is also the main method for pattern making in countries with more developed garment industries.

平面法是服装纸样制作中使用最为广泛的一种方法，它是将服装的立体形态转化为具体的平面展开图，即通过服装把人体的立体三维关系转换成服装纸样的二维关系，并通过定寸或公式绘制出平面的图形（纸样）。平面法得益于板师们长期的经验积累和科学发展，使得它简捷、方便、经济实用、科学有效，而且在服装工业较发达的国家和地区也是以此方法作为纸样设计的主流。

2. The draping method / 立体法

This is an old but current pattern design method. As early as around the 13th century, some countries in Europe have used the three-dimensional method to cut clothes. The three-dimensional method is that designs are made by draping fabric directly around the body or a dress stand and the fabric is constructed to the designed shape by making folds, darts and gathers, and lifting and dragging, and finally flatten out the shaped fabric into be the flat pattern. Folds, curves and asymmetric shapes are easy to be displayed using the draping method but difficult by the flat method. Garments from patterns made by the draping method will create natural and fluent fitted effects. The draping method can help designers express their talent and style. The draping method is efficient for evening wear and dress design, and for light fabrics such as silk, velvet and polyester georgette, etc., but it is not efficient for thick, heavy and hard fabrics.

立体法是一种既古老又年轻的结构设计手法。早在公元 13 世纪前后，欧洲的一些国家已经用立体法来裁剪服装。立体法是将衣料直接披覆在人体或人台上，将布料通过折叠、收省、细褶、提拉等手法做成如效果图所示的设计造型，然后展平成二维的布样。在平面法中较难表现的服装褶皱、曲线等不对称性造型，在立体法中均能被较好地表现出来。用立体法获得纸样制成的服装，贴体合身，衣缝线条自然、流畅。立体法有利于发挥出设计者的才华和风格。立体法适用于女式晚礼服、连衣裙等服装设计，并适合轻薄柔软面料如丝绸、丝绒和涤纶乔其纱等，但对厚重硬挺的面料来说，它则没有优势。

However, the draping method requires good conditions such as a standard dress stand and a skilled pattern maker with artistic education. Compared with the flat pattern method, the draping method needs a large amount of material, which can be costly and more work is needed. Therefore, the draping method is usually applied less in modern garment production but more in advanced fashion making or artistic and performance fashion fields.

但是，运用立体法获取纸样时操作条件要求高，需要标准人台，对操作者的技术素质和艺术修养要求也高，相对平面法而言其材耗大、成本高、效率低，因此，在现代服装工业生产中使用较少，而在高级时装制作或艺术性、表演性强的服装领域中应用较多。

3. The method of combining the flat with draping / 立体和平面相结合法

Nowadays, the method of combining the flat with draping is popular in fashion design and also applied in basic pattern making. This method combines the advantages of the flat and draping method and is a developing trend because it is complementary, high efficient and creates a good effect.

当前，在服装设计中运用立体和平面相结合的方法很普遍，而且也被广泛应用于获取基本纸样。这种方法将平面法与立体法各自的优点结合在一起，能互相弥补缺点，效率高、效果好，是发展的必然趋势。

Unit 2　Japanese Bunka prototype for women

第二节　日本文化式女装原型

This garment prototype is founded in Japan and is a basic Japanese pattern. It is based on the net size of the body. It is the basic garment prototype with the added basic ease (for breathing and activity) after being stretched to a flat pattern. From this, a variety of styles can be created.

服装原型创立于日本，也即日本式的基础纸样。它以人体的净尺寸数值为依据，是将原型平面展开后加入了基本放松量（用于呼吸及活动）的服装基本型。然后以此为基础，可进行各种服装的款式变化。

There are many kinds of prototypes in Japan. They can be sorted into three categories: the Bunka prototype, the Doreme prototype and the draping prototype. The detailed methods and features of these three groups are different, but the garments made from these methods create similar outcomes. Among them, the Bunka prototype is the most popular and is widely used. In July 1999, the Japanese Bunka Fashion College produced a new Bunka prototype because of the changing of body shape.

在日本原型派系很多。它们大体可分成三大体系，即文化式原型、登丽美式原型和立体原型。三种原型的制图方法各异、特点不一，但依据它们做出的服装效果相似。其中又以文化式原型流传最广，应用最广泛。由于人体体型在不断变化，日本文化服装学院于 1999 年 7 月推出新文化式女装原型。

Now British prototypes and American prototypes are also used in the international industry. The American prototype can be sorted into suit prototype, fitted prototype and elastic fabric prototype.

目前国际上常用的服装原型还有英式原型、美式原型等。美式原型按所针对服装种类的不同，还可进一步划分为套装的基本原型、贴身服装的基本原型和弹性面料服装的基本原型等。

The Chinese have a similar body type to the Japanese. That is why the Japanese prototype is widely used and developed rapidly in China. Regardless of body type, the same size prototype can be used as long as the bust is the same size. One size prototype can be used for different garments from underwear to coats when ease and shape are adjusted. It is difficult to master the ease, so it needs to be explored and summarized in practice.

中国人的体型与日本人的体型比较接近，因此，日本原型在中国的适用性很强，应用面广泛，并得到快速发展。对于原型而言，无论何种体型，只要胸围尺寸相同，均可以使用同一个规格的原型。而同一个人的内衣乃至外套大衣也仍可以使用同一规格的原型，只是根据相应款式的需要来决定调整松量的大小和造型。松量大小问题较难掌握，需要在实践中不断摸索和总结。

An introduction into how to draw Japanese women's Bunka prototype is as follows.

下面介绍日本文化式女装原型纸样的绘制。

一、Size table for Japanese women's Bunka prototype for woman / 日本文化式女装原型规格尺寸

The body measurements required to draft Japanese women's Bunka prototype are shown in table 2-2-1 and table 2-2-2:

绘制日本文化式女装原型所需的具体尺寸如表 2-2-1、表 2-2-2 所示：

Table 2-2-1 Measurements for upper body prototype （Unit: cm）
表 2-2-1 上装原型的尺寸 （单位：cm）

Body/ 部位 Measurements/ 规格	Bust(net)/ 胸围 （净）	Nape to waist/ 背长	Waist(net)/ 腰围 （净）	Sleeve length/ 袖长
	82cm	38cm	64cm	52cm

Table 2-2-2 Measurements for skirt prototype　（Unit: cm）
表 2-2-2 原型裙的尺寸 （单位：cm）

Body/ 部位 Measurements/ 规格	Waist(net)/ 腰围 （净）	Hip(net)/ 臀围 （净）	Waist to hip/ 腰 臀深（不含腰头）	Skirt length/ 裙 长（不含腰头）
	64cm	90cm	18cm	60cm

NB/ 注：

① Here the size is 9AR(M) in which the waist is 63 cm; the waist is adjusted to 64 cm for simplifying the calculation. (The prototype in the subsequent content of this book defaults to Japanese Women's Bunka Prototype.)

原型的规格型号取为 9AR（M），腰围 63cm，为便于计算将其调整为 64cm。 （本书中后面所讲的原型都默认为日本文化式女装原型）

② The ease value for the bust is 10 cm, for the waist is 2 cm and for the hip is 4 cm.

胸围放松量为 10cm，腰围放松量为 2cm，臀围放松量为 4cm；

③ In the process of drawing the pattern below, the bust (B), waist (W) and hip (H) in the formula are all net measurements.

在下面纸样绘制的过程中，公式里的胸围（B）、腰围（W）、臀围（H）均为净尺寸。

二、Body prototype drafting / 日本文化式女装原型衣身纸样的绘制

The steps shown in Fig. 2-2-1 and Fig. 2-2-2 are as follows / 图 2-2-1、图 2-2-2 中所示步骤如下：

① Rectangle: length=B/2+5cm=46cm, width= Nape to waist =38cm. The upper end is the top flat line, the lower end is the bottom line, the left is the back centerline and the right is the front centerline.

矩形： 长为 B/2+5cm=46cm，宽为背长 =38cm。上端为上平线，下端为下平线，左端为后中心线，右端为前中心线。

② Bust line: armhole depth from the top flat line down is B/6+7cm=20.7cm; And draw a vertical line from the midpoint of the bust line down to the bottom line.

胸围线： 从上平线向下量取袖窿深即 B/6+7cm=20.7cm，并从胸围线的中点开始向下画一条垂直线到底线。

③ Back width: B/6+4.5=18.2cm.

后背宽：B/6+4.5=18.2cm。

④ Chest: B/6+3=16.7cm.

前胸宽线：B/6+3=16.7cm。

⑤ Back neck width: B/20+2.9=7cm = ◎ .

后领宽（用符号◎表示）： ◎ =B/20+2.9cm=7cm。

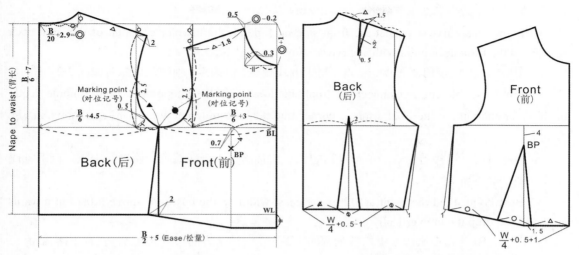

Fig. 2-2-1 Japanese Bunka prototype
图 2-2-1 日本文化式原型

Fig. 2-2-2 Prototype for close ftting
图 2-2-2 腰部合体状态原型

⑥ Back neck depth: it is set to "○"; ○= ◎ /3≈ 2.3cm; draw the back neck curve smoothly from BNP to SNP.

后领深：用符号 ○ 表示；由上平线向上量取后领宽的 1/3 （即 ○= ◎ /3≈2.3cm），再从后颈点到侧颈点画顺后领口弧线。

⑦ Back shoulder line: Use the symbol △ to indicate the length of the back shoulder line. On the back width line, measure ○≈2.3cm downward from the top flat line, then measure 2 cm horizontally to get point SP, and connect it to the side neck point (SNP) in a straight line.

后肩斜线：用符号△表示；在背宽线上由上平线向下量取 ○≈2.3cm，再水平量出 2cm 定点 SP，与侧颈点（SNP）直线连接。

⑧ Back armhole curve: Connect the back shoulder point, the midpoint of the back armhole depth, the end of the bisector of the back armhole Angle, and the midpoint of the bust line, and draw the armhole curve as shown in Fig. 2-2-1.

后袖窿弧线：连接后肩点、后袖窿深中点、后袖窿角平分线端点、胸围线中点，画顺袖窿弧线如图 2-2-1 所示。

⑨ Back waist line: The bottom line.

后腰围线：即下平线。

⑩ Front neck width: ◎ -0.2=6.8cm.

前领宽：◎ -0.2=6.8cm。

⑪ Front neck depth: ◎ +1=8cm.

前领深：◎ +1=8cm。

⑫ Front shoulder line: The side neck point falls 0.5 cm, and measure "2 ◎ /3≈4.7cm" down from the top flat line along the chest width line to get the point, at which make a horizontal line, then draw an oblique line (its length equals to △ -1.8cm) from the side neck point to the horizontal line.

前肩斜线：侧颈点下落 0.5cm，沿胸宽线向下量取距上平线 2 ◎ /3≈4.7cm 处作水平线，从侧颈点向该水平线引斜线（斜线长 = △ -1.8cm）。

⑬ Front neck curve: Connect the front side neck point, the bisector endpoint of the front neck corner and the front neck point with a curve as shown in Fig. 2-2-1.

前领口弧线：连接前侧颈点、前领口角平分线端点、前颈点，画顺领口弧线如图 2-2-1 所示。

⑭ Front armhole curve: Connect the front shoulder point, the midpoint of front armhole depth, the bisector endpoint of the front armhole corner, and the bust midpoint with a curve as shown in Fig. 2-2-1.

前袖窿弧线：连接前肩点、前袖窿深中点、前袖窿角平分线端点、胸围线中点，画顺袖窿弧线如图 2-2-1 所示。

⑮ Bust Point: Find the point at 1/2 of the chest width on the bust line, move it 0.7 cm towards armhole, and drop it 4 cm vertically.

胸高点（BP）：在胸围线上取前胸宽的 1/2 处，向袖窿方向移 0.7cm，再垂直下落 4cm。

⑯ Side line: On the bottom line, get the point which is moving 2 cm away from the midpoint to the direction of the back piece, and then connect this point with the midpoint of the bust line, which is the front and back side seams.

侧缝线：在下平线上，从中点向后片方向移动 2cm 得到一个点，然后将这个点与胸围线中点连接，即前后侧缝线。

⑰ Front waist line: extend the front centerline down by 3.4 cm (half of the front collar width) to make a horizontal line. Draw a vertical line from the bust point down to make the two lines intersect, connecting the intersection point and the side seam point as shown in Fig. 2-2-1.

前腰围线：前中心线向下延长 3.4cm（前领宽的 1/2）处作水平线，从 BP 点向下作垂线使两线相交，连接交点与侧缝点，如图 2-2-1 所示。

⑱ Back shoulder dart, back waist dart and front waist dart: construct these darts as shown in Fig. 2-2-2.

后肩省、后腰省、前腰省的绘制如图 2-2-2 所示。

三、Sleeve prototype drafting / 日本文化式女装原型袖子纸样的绘制

Sleeve prototype drawing is shown in Fig.2-2-3, Fig. 2-2-4.

日本文化式女装原型袖子纸样的绘制如图 2-2-3、图 2-2-4 所示。

Make sure that measure Front Armhole = 20cm and Back Armhole = 20.5cm before drawing sleeves.

绘制袖子前先在前后衣片上量取前 AH=20cm，后 AH=20.5cm。

① Sleeve length line: draw a vertical line as the centerline of the sleeve. The length is the sleeve length (SL= 52cm).

袖中线：作垂直线，长度为袖长 =52cm；

② Sleeve width line: measuring AH/4+2.5=12.6cm down from the top point on sleeve length line, make a horizontal line.

袖宽线：在袖中线上由上向下量取袖山高 =AH/4+2.5=12.6 cm 处，作水平线；

③ Elbow line: measuring SL/2+2.5cm = 28.5cm down from the top point on sleeve length line, make a horizontal line.

袖肘线：在袖中线上由上向下量取袖长 /2+2.5=28.5cm），作水平线；

④ Bottom line: make a horizontal line through the bottom point on sleeve length line.

底边线：过袖中线下端点作水平线；

⑤ Sleeve crown oblique line: draw two fixed-length slant lines from the top of the sleeve crown to the sleeve width line. Length of the left = BAH +1= 21.5cm. Length of the right = FAH=20cm.

袖山斜线：由袖山顶点向袖宽线引 2 条定长斜线，左端长 = 后 AH+1=21.5cm，右端长 = 前 AH=20cm；

⑥ Sleeve side line: from the intersection points of the front or back sleeve crown slant line and sleeve width line, make vertical lines downward respectively, and intersect at the bottom line.

袖底缝线：从前、后袖山斜线与袖宽线的交点开始分别向下作垂直线，分别相交于底边线；

⑦ Sleeve crown curve and sleeve opening curve: first determine the key track points, and then connect the points in turn and draw a curve, as shown in Fig. 2-2-3.

袖山弧线、袖口弧线：先确定关键轨迹点，再依次连接各点并画顺弧线，见图 2-2-3。

⑧ The drawing of one-piece fitted sleeve was done based on the prototype sleeve as shown in Fig. 2-2-4.

一片合体袖的绘制是在原型袖的基础上完成的，如图 2-2-4 所示。

NB: It is important that all "curved measurements" are measured very accurately along the curved line. The sleeve is based on the body prototype to ensure a perfect fit at the armhole.

注意：所有的弧线测量务必精确，这一点非常重要。袖山弧线的依据是原型衣身的袖窿弧线，两者需匹配。

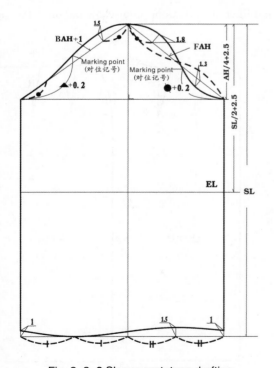

Fig. 2-2-3 Sleeve prototype drafting
图 2-2-3 女装袖子原型绘制

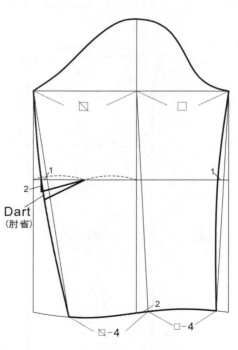

Fig. 2-2-4 Sleeve for close fitting
图 2-2-4 合体袖结构

四、Skirt prototype drafting / 日本文化式女装原型裙子纸样的绘制

Skirt prototype drafting is as shown in Fig.2-2-5, Fig. 2-2-6.

日本文化式女装原型裙子纸样的绘制如图 2-2-5、图 2-2-6 所示。

① Rectangle: Length=60cm; width=H/2+2cm=47cm; the upper line is waistline; the bottom line is hemline; the left line is back centerline, and the right line is front centerline.

矩形：长为裙长 =60cm，宽为 H/2+2cm=47cm，上平线即腰围基础线，下平线即裙下摆线，左端为后中线，右端为前中线。

② Hip line: draw a parallel line from the waistline down, and the distance between them is waist to hip = 18cm.

臀围线：从腰围线向下作平行线，其间距为臀长 =18cm。

③ Hip girth: front hip girth = H/4+1cm (ease) +1cm (difference) = 24.5cm; back hip girth = H/4+1cm (ease) -1cm (difference) = 22.5cm.

臀围：在臀围线上，前臀围 =H/4+1cm（放松量）+1cm（前后差）=24.5cm；后臀围 =H/4+1cm（放松量）-1cm（前后差）=22.5cm。

④ Front waist girth: W/4+0.5 (ease) +1cm (difference)+5cm (dart) = 22.5cm; the waistline should be raised 0.7 cm up at the side seam, and draw the front waistline curve smoothly.

前腰围：W/4+0.5cm（放松量）+1cm（前后差）+5cm（省量）=22.5cm，腰围线在侧缝处起翘 0.7cm，画顺前腰围弧线。

⑤ Back waist girth: W/4+0.5cm (ease) -1cm (difference)+5cm (dart) = 20.5cm; the waistline should be dropped 1.5 cm at the back center line, and lifted 0.7 cm at the side seam, and draw the back

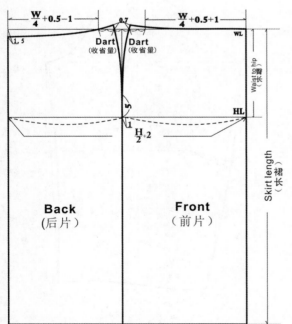

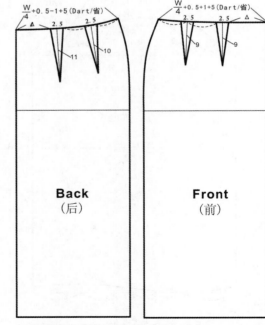

Fig.2-2-5 Skirt prototype drafting
图 2-2-5 裙子原型绘制

Fig.2-2-6 Skirt prototype drafting with darts
图 2-2-6 省道定量的裙原型

waistline curve.

后腰围：W/4+0.5cm（放松量）-1cm（前后差）+5cm（省量）=20.5cm，腰围线在后中心线处下落 1.5cm，在侧缝处起翘 0.7cm，画顺后腰围弧线。

⑥ Side seam: Set a point 5 cm above the hip line. Draw a curve from the point where the front and back side seams are raised to this point respectively, pass the front and back hip boundary points, and then go down until it intersects with the hemline as shown in Fig. 2-2-5.

侧缝线：臀围线向上 5cm 处定一点。分别从前、后侧缝起翘点开始向此点画弧线，经过前、后臀围分界点，再向下，直到相交于下摆线，如图 2-2-5 所示。

⑦ Skirt prototype with darts set according to Thoracolumbar Difference is shown in Fig. 2-2-6.

根据臀腰差定省道的裙子原型绘制如图 2-2-6 所示。

Unit 3 Fitting of Japanese Bunka prototype for women

第三节 日本文化式女装原型试样

一、Shapes of fitting of Japanese Bunka Prototype / 日本文化式原型的试样状态

1. A loosely fitted shape / 衣身呈宽松状态

As shown in Fig. 2-3-1, there are no darts on the body or sleeve prototypes except for the shoulder dart on the back prototype. The garment shapes made from the prototype's pattern can be shown in Fig. 2-3-2. Through the front, back and sides views, you can see that the shoulder and armpit of the garment fit the body very well, but it is loosely fitted on the waist and extended forward and backward to the shape of A. The sleeves is loosely fitted, too. It's defined as a loosely fitted shape.

在图 2-3-1 所示的原型结构图中，除了后衣片保留了肩省外，整个衣身和袖片都不设省道。

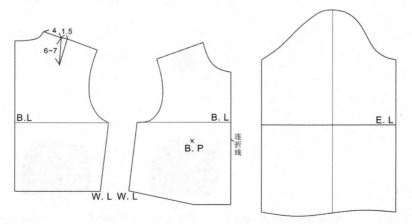

Fig. 2-3-1 Prototypes for a loosely fitted shape

图 2-3-1 衣身呈宽松状态的原型结构图

Fig. 2-3-2 Pictures for a loosely fitted shape
图 2-3-2 衣身呈宽松状态的原型着装效果图

使用该结构图缝制原型上衣，最后生成的着装效果如图 2-3-2 所示，通过正面、背面和侧面的视图，可以看到，衣身的肩部与袖窿部位能够很好地与人体相贴合，但衣身的整个腰部非常宽松，明显出现向前、向后扩展而呈 A 型，袖子亦宽大。可把它看作为宽松状态。

2. A generally fitted shape / 衣身呈一般合体状态

As shown in Fig. 2-3-3, there is a shoulder dart on the back prototype and a dart on the waist seam of the front prototype. There are no waist darts on the back prototype. The bust dart amount is set to "α_1", sufficient to satisfy the breast bulge. The wrist curve on the sleeve prototype is set to the exact cuff size. The garments shape made from the prototype pattern can be seen in Fig. 2-3-4. You can obviously see that it is a different shape from the loose fitted shape. It is semi-fitted on the waist without extending forward or backward. The waist shape created is similar to H. It is neither loosely fitted nor tightly fitted (The sleeve is omitted). It's defined as a generally fitted shape.

在图 2-3-3 所示的原型结构图中，后衣片仍然保留有肩胛省，前片在腰部增设省道，省道量设为 α_1，满足胸部隆起的乳凸量。袖片上的袖口线设为准确的袖口尺寸。使用该结构图缝制原型上衣，最后生成的着装效果如图 2-3-4 所示，通过正面、背面和侧面的着装效果图，可以看到，与前述宽松状态已明显不同的是，前衣身没有了向前、向后扩展开的现象，腰部顺直呈 H 型，既不很宽松，也不很贴体（袖子省略）。可把它看作为一般合体状态。

Fig. 2-3-3 Prototypes for a generally fitted shape
图 2-3-3 衣身呈较一般状态的原型结构图

Fig. 2-3-4 Pictures for a generally fitted Shape
图 2-3-4 衣身呈一般状态的原型着装效果图

3. Tightly fitted shape / 衣身呈完全合体状态

As shown in Fig. 2-3-5, there is a shoulder dart and a waist dart on the back of the prototype. On the basis of Fig. 2-3-3, a waist dart α_2 is added on the waist seam of the front piece, which is the difference between the bust girth and the waist girth. The bust dart α_1 adds the waist dart α_2 to be the full dart α. The sleeve prototype is modified to a one-piece with elbow dart that is tightly fitted. The garment shapes made from the prototype pattern can be seen in Fig. 2-3-6. You can see that the bust and the waist part of the garment are matched well with the body. The sleeve part is slightly tilted forward and fitted well with the shape of the arm. We define this as a tightly fitted shape.

在图 2-3-5 所示的原型结构图中，后衣片除有肩胛省外，腰线上也设有腰省。前衣片在图 2-3-3 的基础上增设了腰省量 α_2，即胸腰差的量。胸省量 α_1 与腰省量 α_2 合在一起成为全省 α。袖片使用的是带有肘省的一片贴体袖。使用该结构图缝制原型上衣，最后生成的着装效果如图 2-3-6 所示，通过正面、背面和侧面的着装效果图，可以看到，整个衣身的胸部、腰部都能够与人体很好地相贴合，袖身也贴近手臂，呈略微自然前倾状，由此可以把它看作是一种完全合体状态。

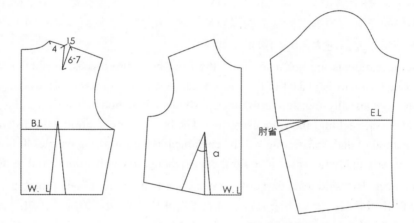

Fig. 2-3-5 Prototypes for a tightly fitted Shape
图 2-3-5 衣身呈完全合体状态的原型结构图

Fig. 2-3-6 Pictures for a tightly fitted shape
图 2-3-6 衣身呈完全合体状态的原型着装效果图

二、Analysis of the fitting for Japanese Bunka Prototype / 原型的试样分析

Daily life garments are usually sorted into three styles: loosely fitted garment, generally fitted garment and tightly fitted garment. They are the same fit as the Japanese Bunka Prototype. Also, the waist ease required is the same.

日常生活服装大致概括为三种造型风格，即宽松服装、一般合体服装、贴体服装。它们正好与前面所述的日本文化原型的三种试样状态相一致，此外对衣身腰部的松量要求亦相吻合。

As shown on the structure graph (Fig. 2-2-2) of the Japanese Bunka Prototype, the dart positioned in the waist on the front body prototype combines the bust dart and the front waist dart. The bust dart value is the ease value of the bust depth and set to α_1. The front waist dart value is the difference between the bust girth and the waist girth and is set to α_2. The value of α_1 is depended on the bust depth. The greater the bust depth, the bigger the value of α_1 is. The value of α_2 is depended on the difference between the bust girth and the waist girth. The greater the deviation, the bigger the value of α_2 is.

在日本文化式女装原型的结构图（图 2-2-2）中，在前衣片腰部的省实质上包含了胸省和前腰省：胸省是用于满足胸部隆起的胸凸量（设为 α_1），前腰省是胸腰差量（设为 α_2）。α_1 的大小主要取决于胸部的丰满程度。胸部越是丰满，α_1 的数值就越大。α_2 的大小主要取决于胸围与腰围的差量。差量越大，α_2 的数值也就越大。

In general, garments are not made from the front body prototypes, as shown in Fig. 2-2-2, in which only one bust dart is placed in the waist because it is not a reasonable way to make garments. The bust dart α_1 is usually transferred to other positions of the body such as the side seam, shoulder line, armhole curve, neckline and front centerline. The bust dart α_2 is left in the waist, as shown in Fig. 2-3-7. The method of dart transferring will be explained in the following chapter. It is emphasized that the bust dart α_1 is actually the ease value for the bust, though it can be transferred to the shoulder dart, armhole dart, neck dart, side seam dart or front center dart on the front body prototype.

在实际制作服装时，一般不采取将所有省量都集中设在腰部的做法（如图 2-2-2 所示），通常是采取将胸省量 α_1 转移到其他部位，如侧缝、肩部、袖窿、领口、门襟等处。而把省量 α_2 留在腰部作腰省，如图 2-3-7 所示。省道转移的具体变化方法将在后面的章节中讲解。需要特别强调的是，虽然胸省量 α_1 可以被转移到前衣身的肩省、袖窿省、领口省、侧缝省、门襟省等，但其实质就是为满足胸部隆起需要的省量。

The value of α_1 or α_2 is related to the ease value of the garment. When $\alpha_1=0$, $\alpha_2=0$ (minimum), the garment shape becomes the loosely fitted shape. When both α_1 and α_2 take the maximum value, the garment shape maybe becomes a tightly fitted shape. When $0 < \alpha_1 <$ maximum and $0 < \alpha_2 <$ maximum, the garment shape becomes a general fitted shape. It can be seen that the different fitted states of the Japanese Bunka prototype can be transformed into each other by adjusting the values of α_1 and α_2.

α_1 与 α_2 的取值大小与服装的松量相关。当 α_1 和 α_2 都取 0 时，服装就呈现为宽松状态（原型试样的第一种状态）；当 α_1 和 α_2 都取最大值时，服装就呈现为贴体服装（原型试样的第三种状态）；当 α_1 和 α_2 的取值介于最小值与最大值之间时，服装就可能呈现为一般合体状态（原型试样的第二种状态）。可见，日本文化式原型的几种不同试样状态是可以通过调节 α_1 与 α_2

的数值大小而相互转化的。

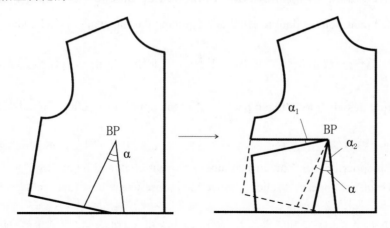

Fig. 2–3–7 Split the dart into two parts
图 2-3-7 把省拆分为两部分

Unit 4 Pattern making for Japanese women's Bunka prototype with CAD software

第四节 用 CAD 软件制作日本文化式女装原型

一、Set size specification / 规格设置

Open "making pattern system" – click "new" and a dialog box of size appears. Double-click a size code, and this size code can be found in the list. Double-click the size code in the list, and it can be deleted. So does name of part.

打开"打板系统"—单击"新建"图标，出现规格设置对话框。双击样板的号型，列表框内就会出现该号型，如果双击列表框内的号型则可以删除。部位名称的操作也一样。

Input measurements of each part after selecting "basic size". Input grading rule if you want to grade. Click "divide all parts" – click "divide" – click "Determine". You can select only one size to get into the pattern making system.

先选择基准码，再输入各部位的尺寸。如果样板需要放码，则输入相应档差，选择"等分所有部位"—单击"等分"，单击"确定"。也可以只选择一个基准尺码，直接进入打板系统。

The basic prototype block of Japanese body measurements are: nape to waist is 38 cm; bust is 84 cm; sleeve length is 53 cm.

日本文化式女装原型规格：背长 38cm，胸围 84cm，袖长 53cm。

The structure of the Japanese Bunka prototype for women can refer to the second unit.

日本文化式女装原型结构图见本章第二节。

二、Pattern making of the front and back bodice blocks / 前、后片衣身纸样制作

(1) Select "Zhizun pen" – double-click and "rectangular" appears – click "start point" – input "47,38" – enter.

选择"智尊笔"—双击出现"矩形"工具，在工作区单击一点作为起点，输入"47，38"—回车。

(2) Right-click to switch to "Zhizun pen" – select the upper line – click – input "21" – enter and get the bust line.

右击切换到"智尊笔"—选择上平线—单击—输入"21"—回车，得到胸围线。

(3) Select the upper line (in Zhizun Pen mode) – right-click and input "18.5" – move the mouse near the left end and "×" appears – click – move the mouse to the bust line – click, and get the back width line.

（在智尊笔模式下）选择上平线—右击，输入"18.5"—鼠标偏向左端时出现"×"—单击—移动鼠标到胸围线—单击，得到背宽线。

(4) In the same way you can get the bust width line, the bisectors of the front and back bodice blocks, the back neck rib and the front neck frame. Make a 2 cm long vertical line from the upper line to the lower 1/3 of the back collar width at the back width line, and get the back shoulder point. From the upper line down to 2/3 of the back neck width, make a vertical line on the chest width line, as shown in Fig. 2-4-1.

同理，可得到胸宽线、前后片平分线、后领深线以及前领框架，从上平线下落 1/3 后领宽处在背宽线上做 2cm 长垂线，得后肩点。从上平线下落 2/3 后领宽处并在胸宽线上做垂线，如图 2-4-1 所示。

(5) Draw the back shoulder line and measure its length, and set it as the parameter "length of back shoulder line = 14.211". Select "Zhizun pen" – click the front neck point – select "projector" – input "length of back shoulder line-1.8" – enter – select the front shoulder horizontal, and click – right-click to end, and get the front shoulder line, as shown in Fig. 2-4-2.

作后肩线并测量其长度，设为参数"后肩线长 =14.211cm"。选择"智尊笔"—单击前颈点—选择"投影点捕捉"工具—输入"后肩线长 -1.8cm"—回车—选择前肩水平线，单击—右击结束，即得前肩线。如图 2-4-2 所示。

(6) Draw the bisector of the bottom corner of the back armhole – input "3" – enter. In the same way, draw the bisector of the bottom corner of the front armhole – input "2.5" – enter.

作后袖底的角平分线—输入长度 3cm—回车。同理，作前袖底的角平分线，输入长度 2.5cm—回车。

(7) On the bottom line, add a point which is 2 cm from the midpoint to the back piece direction, and then connect this point with the midpoint of the bust line, that is, the front and back side seams.

在下平线上，从中点向后片方向移动 2cm 处加一个点，然后将这个点与胸围线中点连接，即前、后侧缝线。

(8) Click "Lengthening or shortening a line" tool, extend the front centerline down at the bottom by "half of the front neck width" – make a horizontal line from the endpoints of the extension to the left – click "vertical line" tool, click "ratio" tool, input "1/2", and make a vertical line from the point half of the chest width downward to the bottom line – click "parallel line" tool, make a parallel line

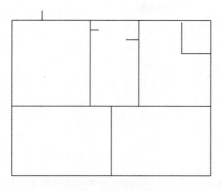

Fig. 2-4-1　The bodice block frame

图 2-4-1　原型框架

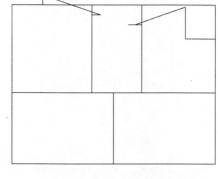

Fig. 2-4-2　Draw shoulder lines

图 2-4-2　作肩线

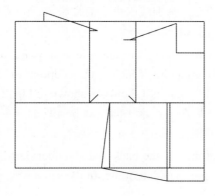

Fig. 2-4-3　Draw the front waistline

图 2-4-3　作前腰围线

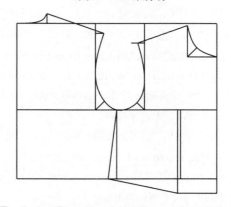

Fig. 2-4-4　Draw the neck curve and armhole curve

图 2-4-4　作领围线、袖窿弧线

of this line 0.7 cm to the left, and extend it downward to meet the horizontal line. Add a point on this line 4 cm below the bust line, which is the bust point. Connect the lower end of the side seam to this intersection – Click the "Break off line" tool to break off the lower horizontal line at the intersection, and delete the excess line segment. As shown in Fig. 2-4-3.

点击"延长或缩短线段"工具，将前中心线在下端向下延长"1/2 前领宽"—在延长线的下端点处向左作一条水平线—点击"垂直线"工具，点击"比例点捕捉"工具，输入"1/2"，从胸宽线一半的点向下作垂线到下平线—点击"平行线"工具，向左 0.7cm 作此线的平行线，并向下延长至与水平线并相交，在此线上从胸围线下 4cm 处加点，即为 BP 点。连接侧缝线下端点和此交点—点击"线段断开"工具，将下端的水平线在交点处断开，删掉多余的线段。如图 2-4-3 所示。

(9) Draw the back neck curve line and adjustment it to the best. Draw the angle bisector of the front neck frame and its length is 1/2 of front neck width minus 0.3 cm. Draw the front neck curve.

作后领口弧线，调整到最佳。作前领框角平分线，长度为"1/2 前领宽 -0.3cm"。画前领口弧线。

(10) Join the back shoulder point, midpoint of back armhole depth, tip of back armhole angle bisector, midpoint of bust line, and the same points on the front piece. You can get the armhole curve. These curves are shown in Fig.2-4-4.

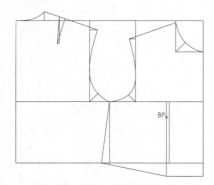
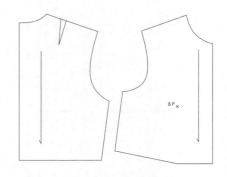

Fig. 2-4-5 Make darts of shoulder
图 2-4-5 作肩省

Fig. 2-4-6 Get patterns of the back and front
图 2-4-6 后衣片、前衣片

连接后肩点、后袖窿深中点、后袖窿角平分线端点、胸围线中点、前袖窿角平分线端点、前袖窿深中点、前肩点，得袖窿弧线。各弧线如图 2-4-4 所示。

(11) Draw a 7 cm long line from the back shoulder line 4cm from the side neck point, and then add a point 0.5cm horizontally away from the left of the endpoint, which is the endpoint of shoulder dart. Draw the two legs of the shoulder dart so that the dart width is 1.5 cm. As shown in Fig. 2-4-5.

从后肩线上距侧颈点 4cm 处画一条 7cm 的线，然后在端点水平向左 0.5cm 处加一个点，即为肩省的省尖点。画出肩省的两条边线，使省宽为 1.5cm。如图 2-4-5 所示。

(12) Use the tool of "take out pattern" to select outlines of the back and shoulder dart line, and right-click to end. Set the pattern name, fabric, number, and then determine. In the same way, you can get the front piece. As shown in Fig. 2-4-6.

利用"样片取出"工具 ，选择后片轮廓线及肩省线，右击结束。设置样片名称及布料、数量，然后确定；同样操作前片。如图 2-4-6 所示。

三、Sleeve pattern making / 袖片制作

(1) Draw a 53 cm long vertical line as the sleeve length line.

作一条长 53cm 的垂线作为袖长线。

(2) Measure the front and back armhole lines, and their total length (AH).

测量前、后袖窿弧线长及总长 AH。

(3) Draw a horizontal line, AH/4+2.5cm away from the sleeve top.

距离袖顶 AH/4+2.5cm 处作水平线。

(4) Draw the front and back sleeve crown slant lines and then a square down from the slant line endpoints to the sleeve opening line. Get the sleeve frame as shown in Fig. 2-4-7.

作前、后袖山斜线，然后从袖山斜线端点至袖口线画一个正方形。得如图 2-4-7 所示框架。

(5) Divide the front sleeve crown slant line into four equal parts. Convex 1.8 cm at the first quarter point and concave 1.5 cm at the third quarter point. Move the second quarter point down 1cm along the slant line. Near the top of sleeve, get a point at 1/4 of front sleeve crown slant line length, and convex 1.8cm on the back sleeve crown slant line. Near the lower end, get a point at 1/4 of front sleeve slant line length plus 1 cm, and Concave 0.5 cm. So you can get the sleeve crown curve, as shown in Fig. 2-4-8.

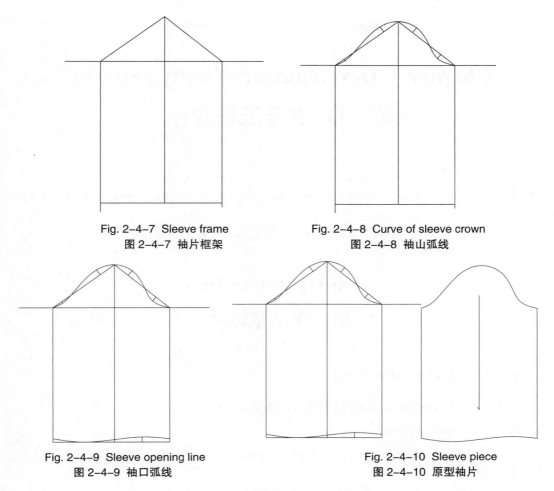

Fig. 2-4-7 Sleeve frame
图 2-4-7 袖片框架

Fig. 2-4-8 Curve of sleeve crown
图 2-4-8 袖山弧线

Fig. 2-4-9 Sleeve opening line
图 2-4-9 袖口弧线

Fig. 2-4-10 Sleeve piece
图 2-4-10 原型袖片

将前袖山斜线四等分。第一等分点外凸 1.8cm，第三等分点内凹 1.5cm，第二等分点沿线下移 1cm；后袖山斜线上，靠顶点一侧取前袖山斜线长的 1/4 处作外凸 1.8cm。靠下端一侧，取前袖山斜线长 /4+1cm 处，作 0.5cm 凹势。得如图 1-4-8 所示袖山弧线。

(6) You also can get the sleeve opening curve according to requirements, as shown in Fig. 2-4-9.
按要求作袖口弧线，得到如图 2-4-9 所示袖口弧线。

(7) Take out sleeve pieces with the "take out piece" tool, as shown in Fig. 2-4-10.
利用 " 样片取出 " 工具 ⬖ ，取出袖片，如图 2-4-10 所示。

Exercise/ 练习

1. Pattern-making for 1:1 scale of Japanese Women's Bunka Prototype by hand (back length 38cm, bust 84cm).
手工绘制 1:1 比例的文化式女装原型图（背长 38cm，胸围 84cm）。

2. Pattern-making for Japanese Women's Bunka Prototype with CAD software.
利用 CAD 软件绘制文化式女装原型图。

Chapter 3 Development of body patterns
第三章 衣身纸样变化

Various shapings around the body can be achieved using darts, gathers, pleats, tucks and styling lines.

通过省道、抽褶、褶裥、塔克和分割线可以塑造各种各样的人体造型。

Unit1 Dart transfer
第一节 省道转移

一、Types of darts / 省道的类型

（一）Darts named by shape / 根据形态命名省道

1. Dart like a nail / 钉子省

This type of dart can be used in expressing the curved surface of the shoulder and breast, such as the neck dart.

这种省道常用于表达肩部和胸部等复杂形态的曲面，如领省。

2. Dart like an awl / 锥子省

This type of dart can be used in expressing tapering surfaces, such as the waist dart or elbow dart.

这种省道用于表达圆锥形曲面，如腰省、肘省。

3. Dart like an olive / 橄榄省

This type of dart can be used in the waist. Such as the waist dart.
这种省道常用于腰部。比如：腰省。

4. Curved dart / 弧形省

This type of dart can be used in extremely fitting fashion, such as the bra.

这种省道常用于极度合体的服装上，比如胸罩。

5. Pleat / 活省

the type of dart has no point，being provided for both functional and decorative purposes.

这种省道没有省尖点，兼备功能性与装饰性。

（二）Darts named by position / 根据位置命名的省道

1. Shoulder dart / 肩省

The root of the dart can be built up into the shoulder line, usually designed like a nail.

省底设在肩线上的省道，肩省常设计成钉子形。

2. Neckline dart / 领口省

The root of the dart can be built up into the neckline, usually designed like an awl.

省底设在领口线上的省道，领口省常设计成锥子形。

3. Armhole dart / 袖窿省

The root of the dart can be built up into the armhole curve, usually designed like an awl.

省底设在袖窿弧线上的省道，袖窿省常设计成锥子形。

4. Underarm dart / 腋下省或侧缝省

The root of the dart can be built up into the side seam, usually designed like an awl on the front of the bodice.

省底设在侧缝线上的省道，侧缝省常在前衣身上设计成锥形。

5. Front center dart / 前中省或门襟省

The root of the dart can be built up into the front centerline, seldom designed like an awl in the front bodice, usually designed like gathers.

省底设在前中线上的省道，很少设计成锥子省，门襟省常设计成碎褶形式。

6. Waist dart / 腰省

The root of the dart can be built up into the waist line. It's usually designed like an awl in skirts and pants, and usually designed like an awl or olive in an upper outer garment and one-piece dress.

省底设在腰线上的省道。在短裙和裤子上腰省常设计成锥形，在上衣和连衣裙上腰省常设计成锥形或橄榄形。

The positions of darts on the bodice are as shown in Fig. 3-1-1.

那些在上衣身上的省道位置如图 3-1-1 所示。

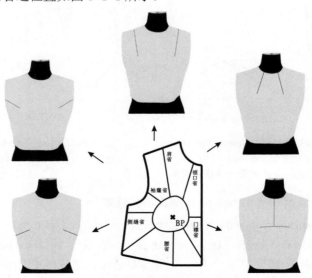

Fig. 3-1-1 The positions of darts
图 3-1-1 省道的位置

二、Theory of dart transfer / 省道转移原理

1. The determinate factors of dart size / 省道大小的决定因素

You may imagine the shape of women's breast as a perfect cone. And that is, folding an angle of y in any direction on a round paper, you can make the round paper become a cone. The angle between the cone edge and the vertical line at the most prominent point of the breast is X (the more prominent the breast, the greater the X value). Even if you fold the paper in different positions, but at the same angle, the cone shape will keep the same, as shown in Fig. 3-1-2.

可把女性胸部隆起而形成的立体形态理想化为一个标准圆锥体。也就是说在一个圆形纸面上任意方向上折叠一个角度为 y 的量，都会使平面的圆变成立体的圆锥体。在胸部最突出点的圆锥体边线与垂直线的夹角为 X（胸部越突出，X 值越大）。虽然折叠的部位不同，但只要折叠的角度相同，其形成的圆锥体的大小和立体效果都保持不变，如图 3-1-2 所示。

On the other hand, assuming that there is another circle with the same center point but a different radius, the dimension of the cone shape is kept the same when the paper circle is folded at the angle y. However, the arc of the circle determined by the angle y is not the same. The longer the radius is, the longer the arc is ($a_2 > a_1$), as shown in Fig. 3-1-3.

另一方面，假设有一个与这个圆的半径不同的同心圆，如果折叠的角度仍然是 Y 的量，那么其圆锥体的立体效果是一样的。但是，由夹角 Y 所确定的弧长是不一样的。半径越长，弧长就越长（$a_2 > a_1$），如图 3-1-3 所示。

The theory of setting darts is the same as that of concentric circles, so the effect of the garment after sewing up the dart is determined by the angle of the dart.

服装设省的理论与同心圆的理论完全相同，因此可以认为服装在省道缝合后的立体效果由省道的角度决定。

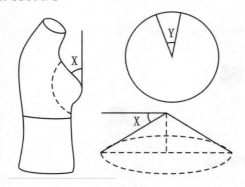

Fig. 3-1-2　The relationship between the shape and the angle of the dart
图 3-1-2　收省后的立体效果与角度的关系

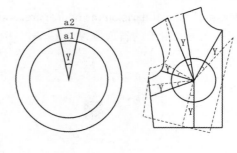

Fig.3-1-3　The size of the dart determined by the angle of the dart
图 3-1-3　省道大小由角度决定

2. The principles of dart transfer / 省道转移的原则

When making patterns from a basic prototype, you often need to transfer darts so that you can get the ideal pattern for variable styles. The principles of transferring dart must be obeyed as follows:

在用原型制板时，因款式千变万化，经常需要进行省道转移才能得到理想的样板。在利用原型进行省道转移时要注意以下几个原则：

(1) The splaying out distance of the new dart is different from that of the old dart, but the dart angles are kept the same. That is to say , the splaying out angles of the new and old darts must be the same, no matter what the new dart position.

省道经转移后，新省道的长度尺寸与原省道的长度尺寸不同，但省道的角度不变，即不论新省道位于何处，新旧省道的张角都必须相等。

(2) If the position of the new dart is not connected to that of the old dart, you should make an auxiliary line passing the bust point so that dart transfer can be done conveniently.

如新省道与原型的省道位置不相接时，应尽量作通过 BP 点的辅助线使两者相接，以便于省道的转移。

(3) No matter how complicated the style is, the dart transfer should keep the balance of the bodice, and keep the waist lines of the front and the back bodices on the same horizontal line (or almost on the same horizontal line), otherwise it will affect the overall balance and measurements of the finished patterns.

无论款式造型怎样复杂，省道的转移要保证衣身的整体平衡，一定要使前、后衣身原型在腰围线处保持在同一水平线上（或基本在同一水平线上），否则会影响制成样板的整体平衡和尺寸的准确性。

三、The principles of dart design / 省道设计原则

1. Dart value / 省量的大小

The dart value can be designed larger according to the body with a large chest and slender waist; the dart value can be designed smaller according to the body with a flat chest and barrel waist. Besides this, the dart value should be designed according to the style. For loose clothing, the dart value should be designed smaller, even with no darts. For fitted clothing, the dart value should be designed larger.

胸部丰满且腰细的体型，省道量可设计得大一点；胸部扁平且腰粗的体型，省道量可设计得小一点。除此之外，设计省道量的大小还应考虑服装的造型风格，宽松服装，省道量应设计的小一些，甚至不设计省道。合体服装，省道量则应设计的大一点。

2. Dart point position / 省尖位置

Dart points usually should fit the body's curved parts at significant points. Chest, scapular, abdominal, buttock and elbow curve of the body are usually considered in clothing, and the significant points of these curves correspond to the bust dart, shoulder dart, stomach dart, waist dart and elbow dart. The significant points have different features, the corresponding dart shape is different. The chest curve is obvious and the bust point position is defined and the dart value is larger. The surface of the scapular curve is large and the significant point is not obvious. This means that the position of the shoulder dart point can be changed within a certain range. The stomach curve and buttock curve are not so noticeable, so that the waist dart can be flexible.

一般省尖点应与人体隆起部位的凸点相吻合。服装中经常考虑人体中的胸凸、肩胛凸、腹凸、臀凸、肘凸，这些部位凸点相对应的省为胸省、肩省、腹省、腰省、肘省。对应不同特征的凸点，省的形状也不同。胸凸明显，胸省省尖位置明确，省量较大。肩胛凸起面积大，无明显高点，故肩胛省的省尖可以在一定的范围内变动。腹部凸起和臀部凸起不那么明显，所以腰部省道设计较为灵活。

As the curved surfaces of the body slope gently rather than sharply, the dart point just needs to point to the position with the greatest changing curvature when the dart is sewn up, and does not need to be sewn into the position. For example, the bust dart on the front piece, its dart point needs to point to the bust point, and when the dart transfer happens, all the transfers need to be done around the bust point at the center. When sewing up the dart, the dart point needs to be kept some distance from the bust point. In general, the shoulder dart point is kept 5–6 cm away from BP point; the armhole dart point is kept 3–4 cm away from the bust point and the underarm dart point is kept 3–4 cm from BP point, the waist dart point is kept 2–3 cm from the bust point, etc. The position of the dart point can be designed freely. In general, the more fitted the clothing is, the closer the dart point is to the curve of the body. On the other hand, the dart point is far away from the curve of the body. By using a bust dart, the clothing becomes fitted, the closer the dart point is to the bust point. For example, the shoulder dart point is kept 2–3 cm from the bust point in fitted clothing.

由于人体曲面变化是平缓而不是突变的，在实际缝制时省端点只要能对准某一曲率变化最大的部位就可以，而不是非得缝制到曲率变化最大的点上。如前衣身上的胸省，其省尖点对准胸高点，在进行省道转移时所有省道转移都以胸点为中心进行。在实际缝制时，省端点距离胸点有一点距离。一般肩省端点距离胸点约 5 ～ 6cm；袖窿省端点距离胸点约 3 ～ 4cm；侧缝省端点距离胸点约 3 ～ 4cm；腰省端点距离胸点约 2 ～ 3cm 等。省尖的位置可以自由设计，但一般来说越是合体的服装，其省道的省尖位置会越靠近人体的凸点，反之越远离人体的凸点。在胸省的使用上就表现为服装越合体，胸省尖点越靠近 BP 点，如合体款式的肩省省尖点距 BP 点也可以取 2 ～ 3cm。

四、The methods of dart transfer / 省道转移的方法

Dart transfer means: any dart can be transferred to any part of the pattern piece around the center point. The fit and measurement of the clothing will be the same after the dart transfer.

省道转移指：服装上某一部位的省道可以围绕着某一中心点被转移到同一衣片上的任何其他部位，同时转移之后不会影响服装的尺寸及合体性。

There are three methods for dart transfer: measuring, rotating and cutting. Each method has its own features. The patterns are based on Japanese women's Bunka prototype and each dart transfer method will be introduced as follows.

省道转移的方法有三种：量取法、旋转法及剪开法，各种方法都有其自身的特点。下面以日本文化式女装原型为基础，来介绍各种省道转移方法。

1. Measuring / 量取法

Measure the side seam lines of the front and back pattern pieces separately, their length difference is the bust dart intake of the front piece. The dart can be drawn out anywhere on the front side line (as shown in Fig. 3-1-4) and the dart point just point at BP. This method is just suitable for the dart whose root can be built up in the side seam.

分别测量前、后衣片侧缝线的长度，其长度差即为前衣身的胸省量。这个省量可以在前侧缝线上腋下任意部位截取（如图 3-1-4），省尖点则指向胸点 BP 点。这种方法仅适用于省道开口在侧缝线上的省道。

As shown in Fig. 3-1-5, first extend the horizontal waist line of the front block of Bunka Prototype for Women, and then draw the side seam lines according to waist specifications directly. After that, measure the length difference between the front and back side seam lines, and it can be anywhere of the front side seam line as the dart intake. Finally, correct the side seam line to keep the two legs of the dart equal and correct the dart point position.

如图 3-1-5 所示，先延长文化式女装原型前腰围的水平线，然后直接根据腰部的规格尺寸作出侧缝线的位置。之后量取前、侧缝线长的差量作为收省量，可在前侧缝线上任意部位截取。最后要注意修正侧缝线使省道的两条边线长度相等，并修正省尖点的位置。

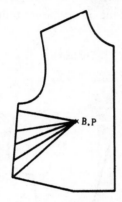
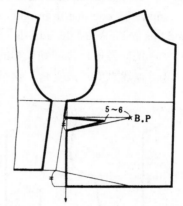

Fig.3-1-4 The directions of underarm darts
图 3-1-4 侧缝省的方向

Fig. 3-1-5 Underarm darts after measuring
图 3-1-5 量取法作侧缝省

2. Rotating / 旋转法

The dart point is at a rotating center (in general, the bust point will be at the rotating center on the front piece), rotate the piece to the desired angle, transfer all or a part of the dart to the other position.

以省端点为旋转中心（一般前片是以衣身的胸点作为样板的旋转中心），旋转衣身一定的角度，将全部省道或省道的一部分转移到其他部位。

Define the bust dart α_1 of Japanese women's Bunka prototype by rotating before explaining the method of transferring dart by rotating, and each step is as follows in Fig. 3-1-6:

讲旋转法转移省道之前，首先要用旋转法定出文化式女装原型上满足胸部隆起的胸省量 α_1，步骤如图 3-1-6 所示：

(1) Trace around the front bodice block, mark out the bust point on the paper, draw a horizontal line from the front waist line and extend the line towards the side seam.

将需要的原型前片样片轮廓描绘在一张画图纸上，在纸上标出 BP 点的位置，并从前片腰线引水平线向侧缝方向延长。

(2) Draw a vertical line from the bust point to the horizontal waist line on the front bodice block required, and the intersection can be marked point A. Copy the line on the paper. The new intersection can be marked A'. The line from the bust point to point A is one of the dart legs.

在原型前片样片上从 BP 点引垂直线与水平腰线相交于点 A，在画图纸上同样复制这条线，交点记为 A'。BP 点到 A 点的连线就是省的一条边线。

(3) With the bust point as the rotating center point, rotate the front bodice block required

counterclockwise until the bottom stop point of the side seam is rotated to the horizontal waist line. Then, draw a line from the bust point to point A' on the front bodice block required. The line intersects the waist line of the front bodice block and marks its intersection as point B. The line from the bust point to point B is the other leg of the dart.

以 BP 点为中心，将原型前片样片按逆时针方向旋转直至侧缝下止点转至在腰围水平线上。然后，从原型前样片上的 BP 点引直线到 A' 点，此线与原型腰线的交点记为 B 点。BP 点到 B 点的连线就是省的另一条边线。

(4) Relax the bust point so that the front bodice returns to its original position. By this time, an angle has been made by one line from point A to the bust point and the other line from the bust point to point B, and this angle is just the bust dart size α_1.

松开 BP 点，使原型前样片回到最初的位置。这时，可以看到原型前样片上 A 点到 BP 点连线与 BP 点到 B 点连线之间形成夹角，这个夹角就是满足胸部隆起的胸省量 α_1。

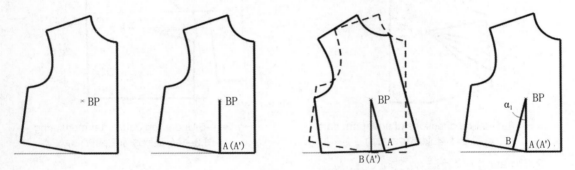

Fig. 3-1-6 Determination of bust dart

图 3-1-6 胸省量的确定

Example 1： Transfer the bust dart α_1 to the underarm dart by rotating (as shown in Fig. 3-1-7).

例 1：用旋转法将胸省量 α_1 转移为侧缝省（如图 3-1-7 所示）。

As shown in Fig. 3-1-8, the steps are as follows:

如图 3-1-8 所示，步骤如下：

(1) Copy the front bodice block as a basic block, and draw the bust dart α_1 on the basic block. The two intersections of the dart legs and the waistline are marked point A and point B respectively.

复制女装原型样板作为基本样板，在基样上作出胸省量 α_1。胸省 α_1 的两条边与腰围线的两个交点分别标为 A 点和 B 点。

(2) Draw a line as the position of the new underarm dart according to the style on the basic block, and this line will be on one leg of the underarm dart, and the new intersection with side seam is marked point C.

根据款式要求在复制的基样上作出新的侧缝省的位置，这条线段是侧缝省的一条边线，与侧缝线的交点记为 C 点。

Fig. 3-1-7 The style with underarm darts

图 3-1-7 有侧缝省的款式

(3) Trace the basic block with heavy lines clockwise from point C (on one leg of the new dart) to point A (on one leg of the original dart) . Then using the bust point as the rotating center, rotate the base block counterclockwise until point B (on the other leg of the original dart) is in the position of point A, and the two legs of the original dart coincide with each other). Point B has now been rotated to the original position of point A, and the position is now known as the position of point B'. Point B has been rotated to the position of point B', and point A have been rotated to the position of point A', and point C have been rotated to the position of point C'.

按顺时针方向用粗实线将点 C（在新省道的一条边上）到点 A（在旧省道的一条边上）的轮廓描下来。以 BP 点为中心，逆时针方向旋转原型样板，直到 B 点旋转到 A 点位置，使旧省的两条边线重合。此时 B 点旋转到 B' 位置（A 点位置），而 A 点则转到了 A' 点的位置，C 点则转到了 C' 的位置。

(4) Trace the basic block with heavy lines counterclockwise from point C' (on one leg of the new dart) to point B' (on one leg of the original dart), and adjust the waist line so that it is horizontal on the paper. By this time, an angle has been made by one line from point C to the bust point and the other line from the bust point to point C'. This angle is just the angle of the underarm dart, which is equal to the bust dart α_1. Finally, adjust the new dart according to the position of the new dart point, which is 3–4 cm away from the bust point to make the two legs of the new dart equal.

按逆时针方向用粗实线将点 C'（在新省道的一条边上）到 B' 点（在旧省道的一条边上）的轮廓（即原型样板的剩余轮廓线）在画图纸上描下来，并修正腰围线使之成为水平线。C 点到 BP 点的连线和 BP 点到 C' 点的连线的夹角即是侧缝省夹角的大小，它与满足胸部隆起的胸省量 α_1 是相等的。最后，重新按照新省尖的位置修正省道，省尖距 BP 点 3 ～ 4cm，注意应该使省道的两边等长。

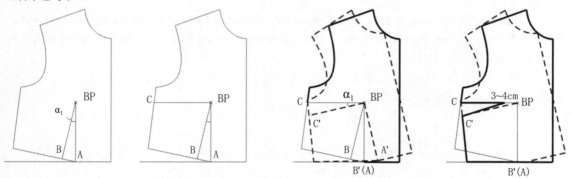

Fig. 3–1–8 Bust dart transferred to underarm dart
图 3-1-8 胸省转移成侧缝省

The method of darts transferred by rotating is simple and convenient and fast. It is suitable for single darts being transferred each other, but it is not suitable for a single dart being transferred into multiple darts. It is not simple when darts need to be transferred to pleats.

用旋转法进行省道转移的方法简单、方便、快捷。对单个省道的相互转移它较为适用，但对单个省变多个省它就不太适用。在需要把省道进行褶裥变化时它也不方便。

3. Cutting / 剪开法

Draw a line according to the new dart position and its style on a basic block. Then cut along the new dart line to the original dart point, fold the original dart along its two legs, and the cut spreads out naturally. The angle of splay is the new dart angle and equal to that of the folded dart.

在复制的基本样板上画出新的省道位置以及新的省道形状，然后沿着新省位线剪开基本样板直至原来的省尖点，沿着原来省道的两边折叠原来的省道，剪开的部位就自然张开。张开的角度就是新省道的角度，它与原先所折叠的省道角度相等。

Example 2: Transfer the bust dart α_1 to the shoulder dart by cutting, as shown in Fig. 3-1-9.

例 2：用剪开法将胸省量 α_1 转移为肩省，如图 3-1-9 所示。

Fig. 3-1-9 The style with
shoulder darts
图 3-1-9 有肩省的款式

As shown in Fig. 3-1-10, the steps are as follows:

如图 3-1-10 所示，步骤如下：

① Copy the front bodice block as a basic block, and draw the bust dart α_1 on the basic block.

复制女装前片原型样板作为基本样板，在基本样板上作出胸省量 α_1。

② Draw a line that points to the bust point to create the new shoulder dart position. This is according to the style.

在复制的基本样板上根据款式在肩线上画出新的肩省位置，省道指向 BP 点。

③ Cut along the new dart line from the shoulder line to the bust point, then fold the original bust dart α_1 along its two legs, and the cut spreads out naturally. The splay angle is the amount of the new shoulder dart and equal to the bust dart α_1.

沿新省道线从肩线开始剪开至 BP 点，然后沿着省道的两条边线折叠原来的胸省 α_1，剪开的部位就自然张开了。张开角的大小即是新省道的量，与原先的胸省量 α_1 是相等的。

④ Trace the developing pattern on the paper with thick lines and mark the bust point. Adjust the new dart to make its two legs equal according to the position of the new dart point, which is kept 5-6 cm away from the bust point.

用粗线将变化的纸样描绘在纸上，标出 BP。根据新的省尖（新省尖距 BP 点 5～6cm）修正省道，使省道两边等长。

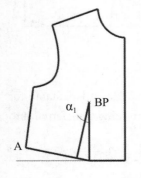
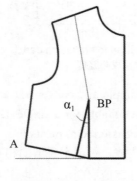
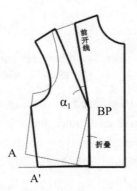
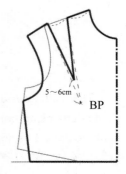

Fig. 3-1-10 Bust dart transferred to shoulder dart
图 3-1-10 胸省转移成肩省

五、The use of dart transfer / 省道转移的应用

（一）The concept of the full bust dart / 全胸省的概念

The combination of the bust dart and waist dart creates the full bust dart α. The full bust dart can be split into two parts (as shown in Fig. 3-1-11). One part is the bust dart $α_1$, and the other part is the waist dart $α_2$. The bust dart $α_1$ can be transferred to the other position by rotating around the bust point, a range of 360°, keeping the effect as the same on the front bodice block. In general, the waist dart $α_2$ is due to the difference between the bust and waist. In theory, the waist dart can not rotate around the bust point and it can only move in the horizontal waist line. The waist dart can change its size according to how fitted the clothing is. The value 0 is the minimum for a waist dart.

全胸省是指胸省和腰省的组合。因此，全胸省可以拆分为胸省和腰省两部分（如图 3-1-11 所示）。一部分是胸省量 $α_1$，另一部分是腰省量 $α_2$。在前衣身上胸省量 $α_1$ 可以通过围绕 BP 点进行旋转（360°范围内），等效转移到其他的部位。通常，腰省量 $α_2$ 是由于胸腰差量引起的，腰省量理论上是不能围绕着 BP 点旋转的，只可以在腰围线上做水平移动。腰省量可以根据服装的合体情况变化，最小时可以为 0。

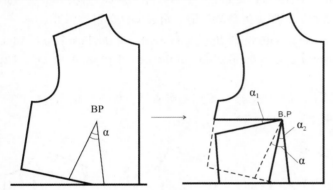

Fig. 3-1-11 Concept of the full bust dart
图 3-1-11 全胸省量的含义

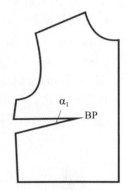

Fig. 3-1-12 Basic block with bust dart in side seams
图 3-12 胸省量在侧缝的基型

To some loose fitting, the bust dart $α_1$ can only be considered to transfer to some positions when making patterns. Therefore, it makes that it's easy for beginners to understand. The bust dart $α_1$ is usually transferred to the side seam to make the waist line horizontal. In this case, the waist dart $α_2$ can be 0. You can consider the front bodice block with the bust dart $α_1$ in side seam as a basic block (as shown in Fig. 3-1-12) to develop patterns in the future study.

对一些腰部并不合体的款式进行结构设计时，经常只需要考虑将胸省量 $α_1$ 进行转移变化。因此，为方便初学者理解，通常把胸省量 $α_1$ 转移到侧缝，以使腰围线成为水平线。这种情况下，可以理解为腰省量 $α_2$ 为 0。在以后的学习中，也可以胸省量 $α_1$ 在侧缝的原型为基本原型(见图 3-1-12) 来进行结构变化。

It needs to consider how clothing is fitted at the waist so that you can decide whether transfer the bust darts $α_1$ or transfer the full bust dart α when making patterns.

在制板过程中，究竟以胸省量 $α_1$ 还是以全省量 α 来进行转移，要根据所设计的服装在腰部贴近人体的情况而定。

（二）The examples of dart developing / 省道变化的实例

1. Transferring the bust dart / 只转移胸省

1）Bust dart transferred to armhole dart / 胸省转移成袖窿省

The style is shown in Fig. 3-1-13. It isn't fitted at the waist, so the waist dart doesn't need to be considered. You do need to consider how to transfer the bust dart α_1 to the armhole.

如图 3-1-13 所示的款式，该款腰部不合身，因此腰部不需要考虑收省。只需考虑将原型上的胸省量 α_1 转移到袖窿部位即可。

Each step shown in Fig. 3-1-14 is as follows / 步骤如图 3-1-14 所示：

① Copy a block having the bust dart α_1 on the side seam, then draw a line which points to the bust point, and it will be the new dart position on the armhole.

复制胸省量 α_1 在侧缝线上的基型为基本样板，然后根据款式在衣片的袖窿处画出新省道的位置，省道指向 BP 点。

② Cut along the new dart line from the armhole to the bust dart point, then fold the original dart along the two legs. The cut of the new dart spreads out naturally.

从袖窿处开始到胸省尖点，剪开新省道，然后折叠原省道，新省道就会自然张开。

③ Trace the developing pattern on a paper with thick lines, and mark out the bust point. Adjust the new dart to make its two legs equal according to the position of the new dart point which is kept 3–4 cm away from the bust point.

用粗线将变化的纸样描绘在一张纸上，标出 BP 点。根据袖窿省的省尖点（距离 BP 点 3 ～ 4cm），修正新省道，以使省道两边等长。

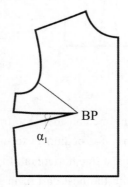
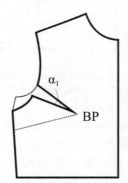
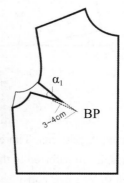

Fig. 3–1–13 The style with armhole darts
图 3-1-13 有袖窿省的款式

Fig. 3–1–14 The bust dart transferred to the armhole dart
图 3-1-14 胸省转移成袖窿省

2）The bust dart transferred to three neck pleats / 胸省转移成领口的三个花省

This is an example of one dart being transferred into multiple darts. The style is shown in Fig. 3-1-15. There are three neck darts in the style. It isn't fitted at the waist, so you don't need to consider the waist dart. You do need to consider how to transfer the bust dart α_1 to three neck pleats. The method of dart transferred by cutting is simple and convenient for this kind of a single dart transferred into multiple darts.

这是单个省转移多个省道的实例。如图 3-1-15 所示的款式，其在领口处有三个开花省，腰部并不合体，不需要考虑腰省。因此只需将胸省量 α_1 转移为三个领口花省即可。这种一个省变多个省的款式用剪开法进行省道转移比较简单、方便。

Each step shown in Fig. 3-1-16 is as follows / 步骤如图 3-1-16 所示：

① Copy a block with the bust dart α_1 in the side seam, then draw three new dart lines LP, MQ and NR on the neck curve according to the style (MQ in the middle of them points to the bust point). After that, draw three auxiliary lines separately to connect the new dart line endpoints P, Q, R with the bust point.

复制胸省量 α_1 在侧缝线上的基型为基本样板，然后根据款式在衣片领口作三个新省位线 LP、MQ、NR（中间的省位线 MQ 指向 BP 点）。之后在基样上作辅助线将新省位线端点与 BP 点相连。

② Cut along the three new dart lines from the neck curve to the bust dart point (be careful not to cut them off), then fold the original dart, and the cuts spread out naturally, and keep the angles of the three new darts splayed out to maintain a balance distribution.

沿新省位线剪开，通过辅助线从颈围线一直剪到胸省端点（注意不要剪断），然后折叠原省道，新省道位会自然张开，并使张开的量被均匀地分配到三个新省中。

③ Trace the developing pattern on a paper with thick line, mark out the bust point. Adjust the three new darts and omit unnecessary lines.

用粗线将变化的纸样描在一张纸上，标出 BP。修正新省道，省略不必要的线条。

 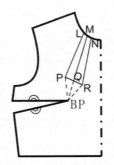 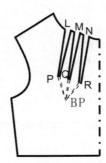 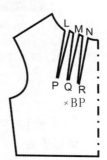

Fig. 3-1-15 The style with three neck pleats
图 3-1-15 领口有三个花省的款式

Fig. 3-1-16 The bust dart transferred to three neck pleats
图 3-1-16 胸省转移成领口的三个花省

2. The full bust dart partly transferred / 全胸省量的部分转移

1）The full bust dart transferred partly to the neck / 全胸省部分转移成领口省

This style is shown in Fig. 3-1-17. It is fitted at the waist with obvious chest curves. It has one neck dart and one waist dart, so you just need to consider how to transfer the bust dart α_1 (a part of the full bust dart α) to the neck dart. The dart α_2 (the rest of that) is the waist dart.

如图 3-1-17 所示的款式，该款式胸部曲线分明，腰部合体。有一个领口省和一个腰省。因此应将全省 α 中满足胸部隆起的胸省量 α_1 转移为图中所示的领口位置，余下的省量 α_2 作为腰省留在腰线上。

Each step shown in Fig. 3-1-18 is as follows / 步骤如图 3-1-18 所示：

① Copy a front bodice block, then draw the full bust dart α and draw a line which points to the bust point as the new dart position on the neck curve line according to the style.

根据款式图在复制的原型前片基样上做出全省 α，并在领口弧线上作指向 BP 点的新省位线。

② Using the BP as the rotating center, rotate the front bodice block counterclockwise until the bottom point of the side seam is on the horizontal waist line. The bust dart $α_1$ (a part of the full bust dart α) is transferred to the neck position by rotation, and the remaining dart $α_2$ is at the waist dart.

以 BP 点为中心逆时针旋转原型，直到侧缝底部与腰线水平，用旋转法将全省 α 中满足胸部隆起的胸省量 $α_1$ 转移到所需的领口位置，余下的省量 $α_2$ 作为腰省留在腰线上。

③ Adjust the positions of the two dart points and make the new dart shape smoothly.

修正省端点位置，修正省道形态，绘制出光滑的省道线。

Fig. 3–1–17 The style with neck
darts and waist darts
图 3–1–17 有领口省和腰省的款式

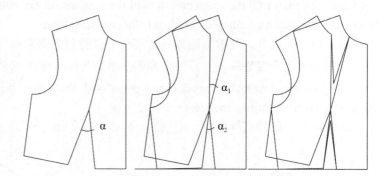

Fig. 3–1–18 Part of the full dart transferred to the neck dart
图 3–1–18 全省部分转移为领省结构变化图

2）The full bust dart transferred partly to two curved shoulder darts / 全胸省部分转移成两个弧形肩省

The style shown in Fig. 3-1-19, is fitted at the waist with obvious chest curves and has two curved shoulder darts and one waist dart. So you only need to consider how to transfer the dart $α_1$ (a part of the full bust dart α) to two curved shoulder darts. The dart $α_2$ (the rest of that) is at the waist dart.

图 3-1-19 中款式胸部曲线分明，腰部合体，有两个弧形肩省和一个腰省。故应将全省 α 中满足胸部隆起的胸省量 $α_1$ 转移为肩部两个弧形省，余下的省量 $α_2$ 作为腰省留在腰线上。

Each step is as follows in Fig. 3-1-20 / 如图 3-1-20 所示步骤如下：

① Copy a front bodice block and draw the full bust dart α. Firstly split the full bust dart α into two parts by rotating. And then one part (the bust dart $α_1$) is transferred to the side seam, and another part is the dart $α_2$ that is kept on waist line.

复制原型前片基样并做出全胸省 α。先用旋转法将全省 α 拆分为两部分。然后把一部分满足胸部隆起的胸省量 $α_1$ 转移到侧缝线上，余下的省量 $α_2$ 作为腰省留在腰线上。

② Copy the developing pattern and draw two curve lines from the shoulder line as the new dart positions according to the style. Then, draw the auxiliary lines to connect the new dart lines with the bust point.

复制变化样板并且在变化样板上根据款式图作出两个弧形肩省的省位线，并作辅助线使之与 BP 点相连。

③ Cut along the two curve dart lines from the shoulder line to the bust dart point passing the auxiliary lines (notice not to cut down), then fold the side seam dart α_1 along the two legs and the two cuts spread out naturally, and keep the angles of the two curved darts splayed out to maintain a balance distribution.

剪开两个弧形肩省的省位线，并沿辅助线剪至 BP 点（不要剪断），折叠侧缝线上胸省量 α_1，两个弧形肩省的省位线处自然张开，使张开的省量均匀分配到两个肩省中。

④ Determine the position of the two curved shoulder darts to make the legs of each dart smooth and equal. Then, adjust the positions of the waist dart and make the new dart shape smooth.

确定两个弧形肩省的省端点使每个省的边线长相等并圆顺，修正腰省尖点位置并使省道形态光滑美观。

Fig. 3-1-19 The style with two curved shoulder darts and waist darts
图 3-1-19 有两个弧形肩省和腰省的款式

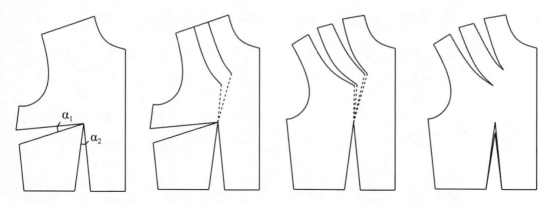

Fig. 3-1-20 Part of the full dart transferred to two curved shoulder darts
图 3-1-20 全省部分转移为两个弧形肩省

3. The full bust dart transfer/ 全胸省量的转移

The full bust dart transfer in asymmetric style / 不对称款式的全胸省量转移

Example 1: Shown in Fig. 3-1-21, the style is asymmetric in which the left dart is on the armhole curve and the right dart is on the side seam.

例 1：如图 3-1-21 所示的款式是左右不对称的。左边省道在袖窿弧线上，右边省道在侧缝线上。

Each step shown in Fig. 3-1-22 is as follows / 如图 3-1-22 所示步骤如下：

① Copy a front bodice block and draw the full bust dart α, and draw a line from the lower part of the armhole curve towards the bust point as a temporary dart position.

复制原型前片基样，在基样上需做出全省（包括满足胸部隆起的胸省量和收腰省的量），从袖窿弧线低位处向胸点画一条线作为省道的临时位置。

② Transfer the full bust dart α to the temporary dart position on the armhole curve by rotating in order not to hamper the dart transfer.

为不妨碍省道的转移，用旋转法将基样上的全省转移至临时袖窿省位。

③ Draw the left and right developing pattern according to the symmetry principle, then draw the two new dart lines according to the style, keeping them touching at Bust point.

Fig. 3-1-21 Asymmetric example 1
图 3-1-21 不对称款式图一

根据左右对称原理，做出变化的左右衣身基样，并按效果图作出新省道线，新省道线指向 BP 点。

④ Cut along the new dart lines to the dart point (notice not to cut down), then fold the armhole dart along the two legs. The cut spreads out naturally. Trace the developing pattern on a paper with thick lines, mark out the bust point and make the waist curve smooth. Adjust the positions of the two dart points and trim the new dart shape smoothly.

沿新省道线剪开至省端点，折叠原省道，新省道线自然张开。用粗线将变化的纸样描在一张纸上，标出 BP 点，把腰围线修顺为光滑弧线。调整省尖点的位置，修顺省的两条边线。

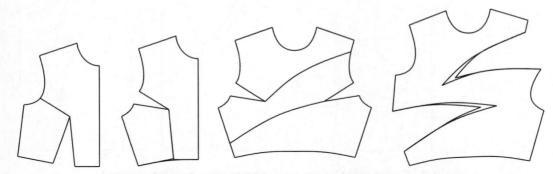

Fig. 3-1-22 Example 1 of the full bust dart transferred to two asymmetric darts
图 3-1-22 不对称全胸省转移款式一

Example 2: The style shown in Fig. 3-1-23 is asymmetric in which the left dart is on the shoulder line and the right shoulder line has no dart.

例 2: 如图 3-1-23 所示的款式是左右不对称的，左边省道在肩线上，右边肩线上没有省道。

Each step shown in Fig. 3-1-24 is as follows / 如图 3-1-24 所示，步骤如下：

① Copy a front bodice block and draw the full bust dart α. Draw a line which points to the bust point as one new dart position in the left shoulder line. Draw another line which points to the bust point according to the style.

复制原型前片基样，并在基样上做出全省 α。在左肩线上作出指向 BP 点的新省道线。根据效果图作出另一条指向 BP 点的新省道线。

② Cut along the new dart line to the dart point, then fold the original dart along the two legs. The cut spreads out naturally. Trace the developing pattern on a paper with thick lines, mark out the bust point and make the waist curve smooth. Adjust the positions of the two dart points and trim the new dart shape smoothly.

沿新省道线剪开至省端点，折叠原省道，剪口自然张开。用粗线将变化的纸样描在一张纸上，标出 BP 点，把腰围线修顺为光滑弧线。修正省尖点的位置，修顺省的两条边线。

Fig. 3-1-23 Asymmetric example 2
图 3-1-23 不对称款式图二

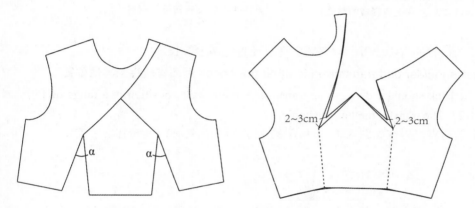

Fig. 3-1-24 The example 2 of the full dart transferred to two asymmetric darts
图 3-1-24 非对称全胸省转移款式二

4. The back shoulder dart transfer / 后肩省的转移

1）The back shoulder dart transferred to the back neck dart / 后肩省转移成后领口省

The style is as shown in Fig. 3-1-25. It has a neck dart on the back piece and is not fitted at the waist so you do not need to consider the waist dart. We just need to consider how to transfer the back shoulder dart to the position on the neck curve.

如图 3-1-25 所示，这个款式在后片领口处有省道，腰部并不合体，因此只需将原型后片上的肩省转移到领口即可，后片腰部并不需要做腰省。

Each step shown in Fig. 3-1-26 is as follows / 如图 3-1-26 所示，步骤如下：

① Copy a back bodice block and draw the shoulder dart on (do not need to draw a waist dart). Draw a line on the back neck curve at the position of the new neck dart to make the neck dart point to the shoulder dart point according to the style.

复制原型后片样板作为基本样板，在复制的样板上做出肩省（腰部不做腰省）。根据款式图，作出后领口省的省位线，尽量使领口省的省尖点指向肩省的省尖点。

② Cut along the new neck dart line to the dart point and fold the shoulder dart. The cut spreads out naturally. Trace around the back bodice and make the shoulder curve smooth. Then adjust the positions of the neck dart points and trim the new dart to make the two dart legs equal.

沿新领口省位线剪开，折叠肩省。剪开处自然张开。描下变化的后片轮廓线，修正肩线为

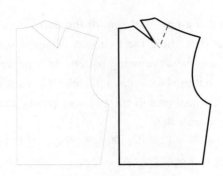

Fig. 3–1–25 The style with
back neck dart
图 3-1-25　后片上有领省的款式

Fig. 3–1–26 The back shoulder dart
transferred to the back neck dart
图 3-1-26　肩省转移为领省

光滑弧线。修正领口省的省尖位置，修正新省道，使省道两边等长。

2）The back shoulder dart transferred to other positions / 后肩省转移到其他位置

Generally, the back shoulder dart can be transferred to any other position in a range of 180°, as shown in Fig.3-1-27. The transfer method is simple.

一般后片上的肩省可以在如图 3-1-27 所示的 180° 范围内转移。其转移方法简单。

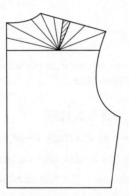

Fig. 3–1–27 The transfer range of the back shoulder dart
图 3-1-27　后片肩省的转移范围

Unit 2 Application of pleats on garments
第二节 褶裥在服装上的应用

Gathers, pleats and tucks are important for garment modeling, and are the structures of clothing that forms a 3-dimensional effect. In order to enrich the styling changes and increase the artistic effects of clothing, you can not only decompose a dart into multiple darts, but also use gathers, pleats and tucks in the clothing structure and other forms of combinations. They are not only to make the garment looser but also to increase some decorative effects, so that the clothing is more artistically appealing.

抽碎褶、打褶裥、做塔克是服装造型中的重要手段，是对服装进行立体处理的结构形式。为了丰富服装的造型变化，增加服装的艺术效果，不但可以将一个省道分解为多个省道，还可以利用服装结构中的抽褶、打褶、做塔克及其他形式的组合来表现。这不但可以增大服装的宽松量，还能增加一些装饰性效果，使服装具有更强的艺术感染力。

一、Classification of pleats / 褶裥的分类

Pleats have a variety of types, which are mainly natural and light gathers, neat folds and free soft tuck pleats. Different pleats have different appearances, which show different styles. Pleats are widely used in women's clothing.

有多种类型，主要有自然轻松的碎褶、整齐利落的褶裥和随意柔和的塔克褶。不同的褶裥有不同的外观形态，也表现出相异的风格特征。褶裥在女装中被广泛应用。

1. Gathers / 碎褶

Gathers can be taken as a combination of many small pleats. Gathers can be transformed from darts and also be designed for a certain decorative effect, but they are more loose and free than a dart. They are often used in women's and children's tops and skirts. The loose dress, as shown in Fig. 3-2-1, shows a transverse dividing line in shoulder which leads to the sleeves, and the style line with continuous gathers, that create the waist loose, and a natural drape effect after tightening the waist belt.

碎褶可以看作由许多细小的褶裥组合而成。碎褶可以由省道转变而来，也可以为了一定的装饰效果而设计，但它比省缝形式宽松、自如。碎褶常常应用在女装和童装的上衣和裙子上。如图 3-2-1 所示的宽松连衣裙，在肩部有横向分割线并一直通向袖子，并且分割线处有连续的碎褶，使得腰部宽松，在束了腰带后呈现出自然的褶皱效果。

2. Pleats / 褶裥

To create a pleat, you must fold one end of the fabric regularly and then fix it by stitching. On the other end of the fabric, you can use a variety of forms, such as stitch fixing, ironing and shaping, or not fixing and spreading out naturally. Pleats are made up of three layers of fabric: the outer, middle and inner layer. The outer layer is a garment piece and covers the middle and the inner layers which become the invisible parts. A deep pleat consists of three layers that

Fig. 3-2-1 A loose dress with gathers

图 3-2-1 有碎褶的宽松连衣裙

are the same fabric size; a shallow pleat consists of three layers of different fabric amounts. Two ruffles of the pleats are visible and hidden. Pleats can be divided into different types due to their different appearance and folding modes.

褶裥就是将面料的一端进行有规则的折叠，并用缝迹固定。而在面料的另一端可以采用多种形式，如用缝迹固定、熨烫定型或是不固定而自然散开等方式。褶裥由三层面料组成，即外层、中层和里层。外层是衣片上的一部分，覆盖着中层和里层，因此中层和里层为不可视部分。由三层同样大小的面料组成的是深褶裥；由三层不同量的面料组成的是浅褶裥。褶裥的两条褶边分别是明褶边和暗褶边。由于外观形态和折叠方式的不同，褶裥可被划分为不同的类型。

1）Classified according to the type of lines forming pleats / 按形成褶裥的线条类型来分类

(1) Straight line pleats / 直线裥

The straight line pleat is that both ends of the pleat have the same value of folding, and its appearance is in parallel lines. It is often used in clothing design such as pleats of pleated skirts.

直线裥指褶裥两端折叠量相同，其外观形成一条条平行的直线，常用于服装的设计如百褶裙上的褶裥。

(2) Curved line pleats / 曲线裥

The curved line pleat is that when the folding value is changing constantly, the pleat has a continuous changing curve in appearance. It is commonly used in the skirt design to match the measurement difference between the human waist and hip. It often contains a dart value, and the pleat folded value is big at the top and small at the bottom of the garment, forming a curve shape appearance. The shape of the curve can only be achieved by sewing to fix pleats on the reverse of the skirt piece.

曲线裥指折叠的量不断变化时，褶裥在外观上形成一条连续变化的弧线。它常用于裙片的设计，以吻合人体腰臀部位的尺寸差异。在褶裥里往往包含了省道的量，并且褶裥的折叠量上大下小，形成弧线形的外观。不过弧线的造型只能通过在裙片的反面车缝固定褶裥来实现。

(3) Oblique tucks / 斜线裥

The oblique tuck is that both ends of the pleat are folded in different ways, but their changes are uniform. In appearance, these pleats form many straight lines that are not parallel to each other. This is commonly used in the skirt design, as shown in Fig.3-2-2.

斜线裥指褶裥两端折叠量不同，但其变化均匀。外观上形成一条条互不平行的直线。它常

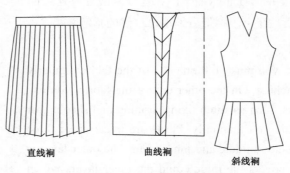

直线裥　　　　　曲线裥　　　　　斜线裥

Fig. 3-2-2 Several forms of pleats

图 3-2-2 褶裥的几种形式

用于裙片的设计，如图所示 3-2-2。

2）Classified according to the appearance and form of pleats / 按形成褶裥的形态来分类

(1) A shun pleat / 一顺裥

A shun pleat is that the folded pleats are pointing in the same direction, also known as the "wind pleats", as shown in Fig. 3-2-3.

一顺裥指向同一方向折叠的裥，亦称顺风裥，如图 3-2-3 所示。

(2) Box pleats / 箱形裥

The Box pleat is that the fabric is folded in two directions at the same time. The two visible pleat edges of the box pleat coincide together to form a hidden pleat (sometimes called the "dark pleat"). The two hidden pleat edges coincide together to form a visible pleat (also called the "light pleat"). As shown in Fig. 3-2-4.

箱形裥是指同时向两个方向折叠的裥。箱形裥的两条明折边重合在一起就形成阴裥（有时又称暗褶）。箱形裥的两条暗折边重合在一起就形成阳裥（又称为明褶）。如图 3-2-4 所示。

(3) Accordion pleats / 风琴裥

The accordion pleat is that there are no folding fabric. Create the pleats only by ironing and shaping. This kind of pleating forms a visible crease and a hidden crease only on the fabric surface, and there are no common three-layer fabric relationship of pleat (as shown in Fig. 3-2-5).

风琴裥指面料没有折叠，只是通过熨烫定型形成褶裥效应。这种褶裥仅仅在面料的表面形成明褶边和暗褶边的褶痕，而没有常见褶裥的三层面料关系（如图 3-2-5 所示）。

 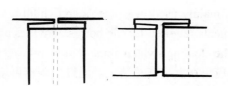 

Fig. 3–2–3 A shun pleat form
图 3–2–3 一顺裥形式

Fig. 3–2–4 Two types of box pleats
图 3–2–4 箱形裥的两种形式

Fig. 3–2–5 A accordion pleat skirt
图 3–2–5 风琴裥裙

3. Tucks / 塔克褶

The tuck pleat is similar in structure to the straight pleat. The same point is that the fabric needs to be folded regularly to one side. And then it is fixed in place with stitching. The difference is that the common pleat just needs to iron the part fixed with stitching, the other part of the pleat open naturally. The tuck pleat has more decorative effects than the common pleat.

塔克褶在结构上类似褶裥。相同之处是都需要将面料有规律地折倒倒向一侧，再用缝迹固定。不同之处在于普通褶仅仅熨烫缝迹固定的一边，其余部位则自然张开。塔克褶比普通褶裥更具有装饰效果。

(1) General tucks / 普通塔克

The general tuck is that the fabric uses stitch fixation along the visible crease of the folded

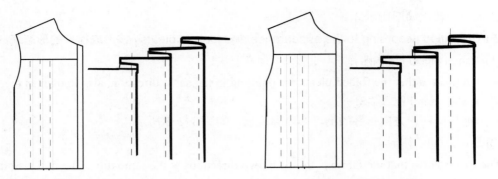

Fig. 3-2-6 General tucks
图 3-2-6 普通塔克

Fig. 3-2-7 Vertical tucks
图 3-2-7 立式塔克

inverted pleat (as shown in Fig. 3-2-6).

普通塔克指在面料上沿折倒褶裥的明折边用缝迹固定（如图 3-2-6 所示）。

(2) Vertical tucks / 立式塔克

The vertical tuck is that the fabric uses stitch fixation along the hidden crease of the folded inverted pleat. Because there is no use of the stitch fixation on the visible crease, the vertical tuck is more three-dimensional and has a more sculptured effect than the general tuck (as shown in Fig. 3-2-7).

立式塔克指在面料上沿折倒褶裥的暗折边用缝迹固定。因为明折边没有用缝迹固定，所以立式塔克比普通塔克更具有立体感、更具浮雕效果（如图 3-2-7 所示）。

二、Design and application of gathers / 碎褶的设计与应用

Gathers: this structure is mainly used in women's and children's wear. Through gathers, it can increase the looseness of garments required for the body's activities and play a role in modifying the body shape, and also change the clothing style, making it more relaxed and youthful.

抽褶这种结构形式主要应用在女装和童装上。通过抽碎褶可增大服装的宽松量，便于人体活动，又可起到修饰体形的作用，还能改变服装的风格，使服装显得轻松活泼，充满青春气息。

1. Non-continuous gathers / 非连续抽褶

1）The waist dart with gathers / 腰省上设计碎褶

The waist dart is designed with gathers, as shown in Fig. 3-2-8. It is the combination of gathers and the waist drat.

在腰省上设计碎褶，如图 3-2-8 所示。它是腰省与碎褶相结合的款式。

Due to the waist being a fitted style, the gathers is combined with the waist dart. Therefore, when copying the prototype , the full bust dart α is made on the waist line of the prototype. This is because the gathering is located in the waist dart. There is no need for a dart transfer, but only need to add the amount of gathers.

由于该款式腰部非常合体，同时碎褶又与原型的腰省相结合，因而复制原型基样时，在基样的腰围线上做出全省 α 。而由于碎褶就设在腰省上，因而不需要进行省的转移，只需加放碎褶量。

Each step shown in Fig. 3-2-9 is as follows / 如图 3-2-9 所示，步骤如下：

① According to the style, make the full bust dart on the copied block of the front bodice.

根据款式图在复制的原型前片基样上做出全省。

② Draw a few equal auxiliary lines to the side seam.

在省的边线上向侧缝线做几条辅助线并均匀分布。

③ Cut along several auxiliary lines horizontally, pay close attention not to cut away (just to keep a little), open the auxiliary lines, add the amount of gathers required, the gathers quantity should also follow the principle of uniform distribution.

沿着几条辅助线向侧缝线方向剪开，注意不要剪断（只保留一点），拉开辅助线，加放所需的褶量，放出的褶量也要遵从均匀分配的原则。

④ Correct the position of the dart point and the curve so that it can be connected smoothly.

修正省尖点的位置并修正弧线，使之光滑连接。

⑤ Finally, mark the symbol which means to gather and mark the grain line.

最后，标注抽褶符号和纱向。

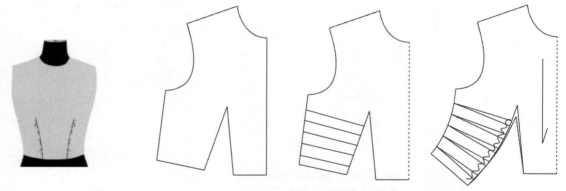

Fig. 3-2-8 Gathers on waist darts
图 3-2-8 腰省上的碎褶

Fig. 3-2-9 The pattern of gathers on waist darts
图 3-2-9 腰省上的碎褶结构图

2）Gathers on the side seam dart / 侧缝省上设计碎褶

As shown in Fig. 3-2-10, this is an example of gathers designed on the bodices side seam dart.

如图 3-2-10 所示的是一款在衣片的侧缝省上设计碎褶的例子。

This style is also very fitted at the waist. When copying the front bodice block, make a full bust dart α on the waistline of the block and the gathers is located on the side seam dart. Therefore, the dart needs to be transferred, then cut the pattern and add the gathers' value.

该款式腰部也非常合体。复制原型基样时，在基样的腰围线上做出全省 α，并且碎褶就设在侧缝省上。因此，需要先进行省的转移，再剪切样板加放出需要的碎褶量。

Each step shown in Fig. 3-2-11 is as follows/ 如图 3-2-11 所示，步骤如下：

① According to the style, make the full bust dart on the copied block of the front bodice and set a side seam dart position. Cut along the dart line, fold the full bust dart and make the curve smooth.

根据款式图在复制的原型前片基样上做出全省，并定出侧缝省位线。沿侧缝省位线剪开，折叠全省，修顺弧线。

② Make several auxiliary lines upward in the side seam dart line.

在侧缝省位线上向上作几条辅助线。

③ Cut along following several auxiliary lines, pay attention not to cut away (just keep a little), open the auxiliary lines and create the desired gathers. Finally, correct to make the curve smooth and mark the symbol of gathers and the grain line.

沿着几条辅助线向上剪开，注意不要剪断（只保留一点），拉开辅助线，补足所需的褶量。修顺弧线，使之光滑连接，并且标注抽褶符号和纱向。

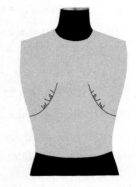

Fig. 3-2-10 Gathers on the side seam dart
图 3-2-10 侧缝省上的碎褶

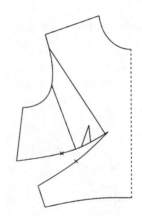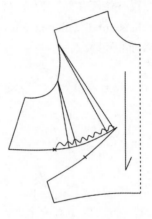

Fig. 3-2-11 The pattern of gathers on the side seam dart
图 3-2-11 侧缝省上的碎褶结构图

2. Continuous gathers/ 连续抽褶

Gathers are designed on the styling curve line on the front of the body ／ 前衣身的弧形分割线上设计碎褶

The style shown in Fig. 3-2-12 is an example of gathers design on the styling curve line of the front of the body.

如图 3-2-12 所示的款式是在前衣身的弧形分割线上设计碎褶的例子。

This style is not very fitted at the waist. When copying the front bodice block, only make the bust dart α_1 on the waistline of the block, which is sufficient for the needs of the bust size. Bust dart α_1 is transferred to the styling curve line as the gathering quantity. There is still a need for an auxiliary line from the styling curve line to the waist line to make up the amount of pleating because the pleat is not up to the designed amount.

该款式腰部并不合体，因而复制原型基样时，在基样的腰围线上只需做出满足胸部隆起的胸省 α_1，并将其转移到弧形分割线上作为碎褶量，因该褶量达不到款式图上所需的褶量，还需从分割线处向腰线方向作辅助线以便补足抽褶量。

Each step shown in Fig. 3-2-13 is as follows / 如图 3-2-13 所示步骤如下：

① Make the bust dart α_1 on the front bodice block of the copied prototype according to the style, and draw a curve style line, then draw an auxiliary line from the bust point to the curve style line.

按图在复制的原型前片基样上做出胸省 α_1 和弧形分割线；并从 BP 点作一条辅助线到分割线上。

② Cut the block along the styling curve line, and cut it to the bust point along the auxiliary line. Fold the bust dart, then transfer the dart to the pleated place and correct the curve line.

沿分割线剪开基样，并沿辅助线剪到 BP 点。折叠基样省道，将省道转移至抽褶处，修正需抽褶的分割线。

Fig. 3-2-12 Gathers on the styling curve line
图 3-2-12 前衣身弧形分割线上的碎褶

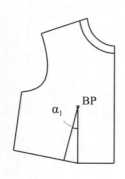   

Fig. 3-2-13 The pattern of gathers on the styling curve line
图 3-2-13 前衣身弧形分割线上的碎褶结构图

③ Because the amount of gathers is large, the bust dart must be transferred to the pleats, but its intake is not enough. Therefore, auxiliary lines should be made along the gathering direction, so as to cut the pattern and make up the required amount of gathers.

由于与分割线连接的碎褶量较多，仅仅利用胸省转移得到的褶量很少，所以还需沿着碎褶的方向做辅助线，以便剪切样板，补足所需的碎褶量。

④ Cut along the auxiliary line and open the cut to make up the required amount of gathers.

剪开辅助线，拉开切口以补足所需抽褶量。

⑤ Correct the waist line and the styling curve line that need gathers.

修正腰围线和需抽褶的分割线。

3. Gathers instead of dart / 代替省道的碎褶

Transfer the full bust dart into gathers / 全胸省全部转移成碎褶

As shown in Fig. 3-2-14, this style is very fitted at the waist. When copying the front bodice block, make the full bust dart on the waistline of the block and transfer it into gathers, as shown in the figure. There is no need to increase the pleats, as there is little need for pleats in this style.

如图 3-2-14 所示，该款式腰部非常合体，因而复制原型基样时，在基样的腰围线上要作出全省，并将其转移为碎褶，因款式中要求的褶量不多，因而就不需再加大褶量。

Each step shown in Fig. 3-2-15 is as follows / 如图 3-2-15 所示，步骤如下：

① According to the style, make the full bust dart on the copied block of the front bodice, and also draw a styling line across the bust and an auxiliary line from the bust point to the styling line.

根据款式在复制的原型前片基样上做出全省，作胸部横向弧形分割线以及一条从 BP 点到分割线上的辅助线。

② Cut the block along the styling line and cut it along the auxiliary line to the bust point; fold the full bust dart and transfer the dart value to the pleating.

沿分割线剪开基样，并沿辅助线剪到胸点；折叠基样省道，将省道量转移至抽褶处。

③ Correct the styling line and waistline, making them into smooth curves. The transferred dart value is for making small gathers.

修正需打褶的弧线和腰线为光滑的弧线。用转移来的省道量抽碎褶。

Fig. 3-2-14　Gathers instead of
the full bust dart
图 3-2-14 代替前胸省道的抽褶

Fig. 3-2-15　The pattern of gathers instead of the full bust dart
图 3-2-15 代替前胸省道的抽褶结构图

4. Decorative gathers / 装饰抽褶

Decorate gathers on the back transverse styling line / 后衣身横向分割线上装饰碎褶

The style as shown in Fig. 3-2-16 is an example of decorating gathers on the back transverse styling line.

如图 3-2-16 所示的款式是后衣身横向分割线上装饰碎褶的例子。

The pleated portion of the style does not contain a dart, and the waist is also not fitted. So there is no need to make the waist dart while copying the back prototype of bodice. Just close the shoulder dart and transfer it to the transverse styling line, then the gathers value required is added by making auxiliary lines from top to bottom and cut along them and open the cuts.

该款式的抽褶部位上并不包含省道，腰部也不合体。因此复制原型后片基样时不需作出腰省。只需关闭肩省并把它转移到横向分割线处，然后将所需碎褶量通过从上至下作辅助线并剪开拉开辅助线来放出。

Each step shown in Fig. 3-2-17 is as follows / 如图 3-2-17 所示，步骤如下：

① According to the style, make the shoulder dart and transverse styling line on the copy block of the back prototype, and extend the shoulder dart to the styling line to intersect.

按图在复制的原型后片基样上作出肩省和横向分割线，使肩省线延长至与分割线相交。

② Cut along the transverse styling line and make it become two parts. In the upper part fold shoulder dart with the styling line upward and correct the styling line to be curve. In the lower part, create the auxiliary vertical line from the styling line to the waist line (can be single or multiple).

沿横向分割线剪开，分为上下两部分。上部折叠肩省，分割线向上翘起，修正分割线为弧线；下部从分割线处向腰线作垂直辅助线（可以是单条也可以是多条）。

③ Cut along the auxiliary line in the lower part and open the cut horizontally, then add the pleat volume.

在下部样片上剪开辅助线（剪断）并水平拉开，放出抽褶量。

From the above examples, you can see that in most models, the pleat volume designed to be transferred from the dart is not enough. Often it need to cut the pattern to increase the pleat volume.

从上面一些例子可以看出，在作碎褶结构设计时，在多数的款式中由省量转移而来的褶量大小是不够的，往往还需要剪切样板加大褶量。

Fig. 3–2–16 Decorating gathers on the back transverse styling line
图 3-2-16 后衣身横向分割线上装饰碎褶

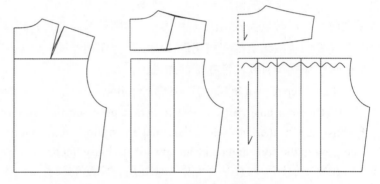

Fig. 3–2–17 Patterns of decorating gathers on the back transverse styling line
图 3-2-17 后衣身横向分割线上装饰碎褶结构图

Commonly there are two ways to increase the pleat volume / 常见的加大褶量的方法有两种：

(1) You can make an auxiliary line as the cutting line to cut along the line and stretch the pattern to add the pleat value.

可作一条辅助线作为剪切线，沿该线剪开样板并拉开以补足抽褶量。

(2) You can make a set of parallel auxiliary lines. Usually, cut the pleated part of the garment along these auxiliary lines and evenly open the cuts to add the pleat value.

可作一组平行的辅助线。通常沿这些辅助线切开衣片抽褶部位，并均匀地拉开切口，以补足抽褶量。

(3) In addition, it is necessary to pay attention, so you should scan patterns carefully to decide

whether it is necessary to cut along the auxiliary line or not. The principle is to pay attention to ensure the overall balance of the front and back bodies. While creating the auxiliary line, you must consider that it maintains in the same direction as far as possible along with the extending pleats.

另外，还须注意的一点就是需不需要剪断辅助线要仔细审视款式图来定，原则是一定要注意保证前、后衣身结构的整体平衡。而作辅助线时，辅助线的方向应尽量和褶的延伸方向一致。

三、Design and application of pleats / 褶裥的设计与应用

1. Application of regular pleats on the front garment / 顺风褶在前衣片上的应用

The style shown in Fig. 3-2-18 is designed with three longitudinal pleats on the front garment.

如图 3-2-18 所示的款式在前衣身上设计了三个纵向褶裥。

The front shoulder of this style has a styling curve line and the lower part of the styling curve line has 3 longitudinal pleats. The waist is not fitted; there is only bust dart to meet the chest uplift on the copied front prototype. Transfer it to the three longitudinal pleats. Because there are three pleats from the styling curve line to the waist line, it is necessary to create three auxiliary lines to add to the pleat quantity. We should pay attention to the portion of the transferred pleating quantity as it should be added to the upper part whilst pleating.

该款式前衣身上肩部有弧形分割线，分割线下部有 3 个纵向褶裥。腰部并不合体，在复制原型前片基样时只需做出满足胸部隆起的胸省量，并将其转移至三个纵向的褶裥当中。因为从分割线处到腰围线上都有褶裥量，所以还须作辅助线，以便加放褶裥量。经转移而来的一部分褶裥量在作褶裥时要注意在上部加进去。

Each step shown in Fig. 3-2-19 is as follows/ 如图 3-2-19 所示，步骤如下：

① Create a bust dart to meet the chest uplift on the copy of the front prototype. First transfer the bust dart to the side seam, then make a styling curve line (the styling curve line length is ＊) according to the style and make an upward auxiliary line from the bust point to the styling curve line.

在复制的原型前片基样上做出满足胸部隆起的胸省量。先把胸省转移到侧缝处，然后根据款式图作出分割线（分割线长定为＊），并从 BP 点向上作一条辅助线至分割线处。

② According to the styling curve line, cut the copied piece along the auxiliary line to the bust point, and fold the original dart. The auxiliary line cut will spread out naturally. Correct the curve（curve length is ＃）.

按分割线，沿辅助线将基样剪开至 BP 点，折叠原省道。辅助线切口处自然张开。修正弧线（弧线长为＃）。

③ Make three auxiliary lines towards the waist on the curve line where the length is ＃, namely the tuck position.

在长为＃的弧线上向腰部作 3 条辅助线，即为褶裥位。

④ Cut along the three auxiliary lines; spread the three cuts for required volumes throughout the pleating at waist level. The difference between the curves ＃ and ＊ should be equally distributed to the parts above the bust point of 3 pleats, and fix

Fig. 3-2-18 The style with three longitudinal pleats on the front piece

图 3-2-18 前衣身有 3 个纵向褶裥的款式

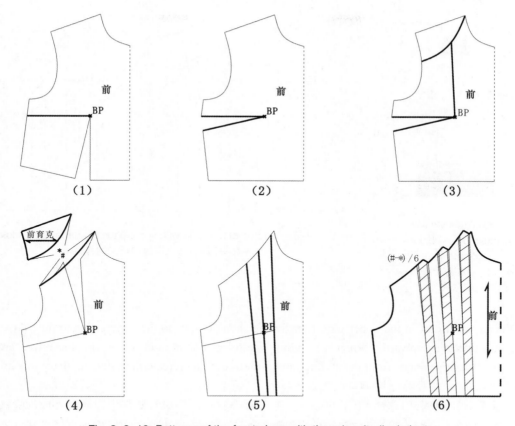

Fig. 3-2-19 Patterns of the front piece with three longitudinal pleats

图 3-2-19 前衣身有 3 个纵向褶裥的结构图

the line. (note: the three pleats are not equivalent from top to bottom)

沿 3 条辅助线剪开，在腰线水平线上拉开各处褶裥所需量，弧线 # 与 * 的差值被平均分配到 3 个褶裥的胸点以上部位。修正连线。（注意：这 3 个褶裥从上至下并不是等量）

2. The application of the tuck pleats on the back bodice of clothing/ 塔克褶在后衣片上的应用

The style shown in Fig.3-2-20 is designed with two vertical tuck pleats on the back bodice of clothing. The methods of changing the tuck structure and pleat process are actually the same, but only use stitch fixing on the pleated part.

如图 3-2-20 所示款式在后衣身上设计两个纵向塔克褶。塔克的结构变化方法实际上和褶裥的处理方法一样，只要在打裥的部位用缝迹固定即可。

Each step shown in Fig. 3-2-21 is as follows/ 如图 3-2-21 所示，步骤如下：

① According to the style, make the shoulder dart and the transverse styling line on the copy of the back of the prototype, extend the shoulder dart to the styling line to intersect.

根据款式图在复制的原型后片基样上作出肩省和横向分割线，将肩省线延长至与分割线相交。

② Cut along the transverse styling line and make it become two parts, the upper and the lower.

Fig.3-2-20 The style with vertical tucks on the back bodice

图 3-2-20 后衣身上有纵向塔克的款式

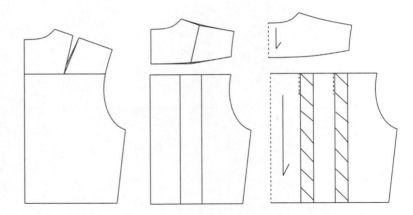

Fig. 3-2-21 Patterns of the style with vertical tucks on the back bodice

图 3-2-21 后衣身上有纵向塔克的样板

Fold the shoulder dart in the upper part, transfer the shoulder dart to the styling line to hide the dart, the styling line goes upward, correct the styling line to be a smooth curve. As for the lower, create two vertical auxiliary lines from the styling line to the waist line (create the same auxiliary line as the tuck).

沿横向分割线剪开，分为上下两部分。上部折叠肩省，将肩省转移到分割线处成为隐藏的省道，分割线向上翘起，修正分割线为光滑弧线；下部从分割线处向腰线作垂直辅助线（有几条塔克线做几条辅助线）。

③ In the lower part, cut along the auxiliary lines and spread the cut horizontally to get the pleat volume, and then use the stitch to fix the pleats to obtain the tuck.

在下部样片上沿辅助线剪开并水平拉开，放出褶裥量，然后用缝迹将褶裥固定即成为塔克。

Unit 3　Design and Application of style lines in women's clothing

第三节　女装上分割线的设计与应用

With its unique directional and flowing characteristics, the line enriches the clothing. Besides darts and pleats, style lines (including seams) are most commonly used in the structure. Style lines can help in creating more fitted clothing than darts or pleats do. Style lines have a variety of forms, such as longitudinal styling line, horizontal styling line, oblique styling line, free styling line and so on. In addition, lines with rhythm and melody are often used, such as radiation and spiral lines.

线条特有的方向性和运动感，赋予了服装丰富的内容和表现力。除了省道、褶裥外，分割线（包括衣缝）是服装设计中最常用的结构形式。它们能设计出比省道和褶裥形式更加合体的服装。服装分割线有各种各样的形态，有纵向分割线、横向分割线、斜向分割线、自由分割线等；此外还常采用具有节奏旋律的线条，如辐射线、螺旋线等。

一、Categories of Style lines / 分割线的分类

Style lines help change the shape of clothing and also can be functional. They have a dominant influence over the shape and fit. In ladies' clothing, mostly curved style lines are adopted to express the feminine form. The shape usually created is a curved waist to highlight the women's physique. showing lively, beautiful, slender lasting appeal. While men's clothing majorly applies straight waist with vertical style lines, no matter how many changes there will be. Considering the characteristics of the style lines, normally, they can be categorized into shape and function lines.

服装上的分割线，既具有造型特点，也有功能作用，它对服装的造型和合体性能起着主导作用。女式服装上大多采用曲线形的分割线，以表现女性的柔美。外形轮廓线以曲腰式为多，显示出活泼、秀丽、苗条的韵味。但是男式服装上的线条无论怎样变化，刚健、豪放的竖直线组成服装的主旋律，外形轮廓以直腰式为多。根据分割线的特点，常将分割线分为两大类：造型分割线和功能分割线。

1. Style lines for shape / 造型分割线

Shape style lines refer to those lines applied to clothing for decorative purposes. They do not help in improving the fitting, but their position, form and quantity changes will affect the artistic feeling of the clothing.

造型分割线是指为了造型的需要而附加在服装上起装饰作用的分割线。它对服装合体与否不起作用，但分割线所处部位、形态、数量的改变会影响服装造型艺术效果。

A single shape style line at a certain position has a limited decorative effect, thus a number of style lines are necessary to get a comparatively perfect shape or to meet the need of the special shape requested. But a balance problem among these style lines will easily occur if the quantity goes up too much. Therefore, the increase in the number of style lines must maintain the overall balance of the clothing to meet certain aesthetic requirements, like rhythm, etc. Especially for horizontal stylelines, the longitudinal position must conform to, or at least close to the golden ratio of 1:1.618, to achieve traditional aesthetic requirements. As shown in Fig. 3-3-1, divided by the bust line and waist line, the upper, middle and lower sections of the jacket are in a ratio of 5:8:5. You can find similar segmentation ratios with other style lines; Fig. 3-3-2 is another example.

单个的造型分割线在某部位上所起的装饰作用是有限的，为了塑造较完美的造型及某些特殊造型的需要，增添分割线是必要的。但分割线数量的增加易引起分割线的配置失去平衡，因此，分割线数量的增加必须保持服装整体的平衡和符合一定的美感要求，比如有一定的节奏感、韵律感等。特别是对于水平分割线，则其纵向位置必须符合黄金分割比率（1：1.618）或接近黄金分割率，才会符合传统的审美要求。如图 3-3-1 所示，经胸围线和腰围线分割，上衣的上、中、下分段比例为 5：8：5。其他分割的比例也相似，如图 3-3-2 所示。

2. Functional style lines / 功能分割线

Functional style lines refer to those lines applied to clothing for a fitted shape, easy to process and with technical features.

功能分割线是指具有适合人体体型、易于加工、具有工艺特征的分割线。

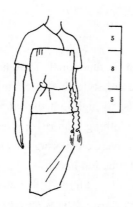
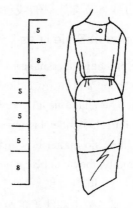

Fig.3-3-1 The style with chest/waist segmentation
图 3-3-1 胸部和腰围有分割线的款式

Fig.3-3-2 The style with top/skirt segmentation
图 3-3-2 上身和裙子有分割的款式

The style lines on clothing are not only for designing a beautiful and novel shape, but also for diversified practical functions, such as making the clothes fitted, highlighting the chest, waist and hip position. It can express the beauty of the human body's curve. Additionally, considering garment sewing requirements, the designed style lines need to mostly ease the complexity of garment processing.

服装上的分割线不仅是为了设计美观新颖的造型，也是为了实现多种实用功能，比如使服装合体，突出胸部、收紧腰部、扩大臀部。这样才能使服装显示出人体的曲线美。此外，考虑到服装的缝制加工方面的要求，设计的分割线需尽可能地减少成衣加工的复杂程度。

Functional style lines have two important characteristics. Firstly, to fit the body well, style lines are simply adopted to show the important curved surface of the body; Secondly, style lines can replace the complicated steaming process of shaping technology and also has or replaces the function of dart. The most typical example is the princess line design. The position of the style line is located in the part where the curvature of the chest changes the most, the shoulder dart is replaced above the chest, and the waist dart is replaced below the chest. The style lines draw the outline of the complicated chest, waist, and hip shape only with simple lines, not only beautifying the shape, but also simplifying the sewing process, making the steaming process unnecessary. Style lines actually function as closing a dart; therefore, they are usually defined as "connecting darts as a seam". (This section deals with functional style line of this kind.)

功能分割线有两个重要的特征：其一，是为了适合人体体型，以简单的分割线形式，最大限度地显示出人体廓线的重要曲面形态；其二，是以简单的分割线形式取代复杂的湿热塑型工艺，兼有或取代省道的作用。最典型的例子，就是公主线的设计，其分割线的位置位于胸部曲率变化最大的部位，胸部以上取代肩省，以下取代腰省，用简单的分割线就把人体复杂的胸、腰、臀部形态描绘出来了，不仅美化了造型，而且简化了缝制工艺，不需要用复杂的湿热塑型工艺来定型。这种分割线实际上起到了收省的作用，通常被称为连省成缝。（在这节里涉及的就是这种具有功能特性的分割线）

Excessive darts not only affect the clothing appearance, but also influence the sewing efficiency, which consequently impacts the durability of the clothing. Therefore, based on maintaining the shape, darts are frequently replaced by sewing seams. That is, connecting these darts into seams or style lines, defined as "connecting darts as a seam".

在服装衣片上作过多的省道，一则影响制品的外观，二则影响制品的缝制效率，从而影响制品的穿着牢度。在不影响款式造型的基础上，常将相关连的省道用衣缝来代替，即将相互关联的省道联合成衣缝或分割线，俗称称连省成缝。

Connecting darts as a seam usually has two forms: seams and style lines, and style lines majorly adopted. Sewing seams mainly include the side seams and back seams, and the rest are collectively referred to as the style lines, such as princess line, knife-back line, etc.

连省成缝的形式主要有衣缝和分割线两种形式，其中以分割线形式占多数。衣缝的缝道形式主要有侧缝、背缝等，其余的统称为分割线，如有公主线、刀背缝线等。

Principles of connecting darts as a seam / 连省成缝的几条原则：

(1) When connecting the darts, the position should be fixed through or close to the body curve point to make the best of the fitted effect of the darts.

省道在连接时，应尽量考虑连接线要通过或接近人体的凹凸变化的点，以充分发挥省道的合体作用。

(2) In the case of connecting between the warp and weft direction darts, from the workmanship point of view, the shortest path connection should be adopted, meanwhile making it easy to process, a good fit and a beautiful shape. From the art point of view, the connection path shall serve for the coordination and unity of the clothing shape. (Here mainly refers to the seam or the form of the style line can not be simply connected in a straight line form, but also from Artistic point of view to consider its beauty and the overall shape of the coordination and unity).

当经向和纬向的省道在连接时，一般从工艺角度考虑，以最短路径连接，并使其具有良好的可加工性，贴体功能性和美观的艺术造型；从艺术角度考虑时，省道相连的路径要服从造型的整体协调和统一。（这里主要是指衣缝或分割线的形态不能简单地以直线形式相连，还应从艺术角度考虑其美观性及与整体造型的协调、统一）。

(3) When connecting the darts as a seam, connection lines need correct in details to make the seams smooth and beautiful, not confined to the original shape of the darts.

省道在连接成缝时，应对连接线进行细部修正，使缝线光滑美观，而不必拘泥于省道原来的形状。

(4) If the connection of the dart is not ideal according to the original shape, the dart should be transferred first and then connected. Pay attention to the dart, so that it does point to the original process point after the transfer.

省道的连接如按原来的形状连接不理想时，应先进行省道的转移再进行连接，但须注意转移后的省道应指向原先的工艺点。

二、Style line constructional design and application / 分割线的结构设计及应用

Seam is also one type of style line. Thus the change method of the style lines are similar to that of connection of darts as a seam. This can be divided into two categories.

因为衣缝本身属于分割线一类的，所以分割线的变化方法与连省成缝处理方法类同。可分

为两大类。

（一）Style lines through convex points of the body / 通过人体凸点的分割线

Style lines, especially functional ones, have more applications on lady's clothing. Moreover, the change of a style line has a major focus on the application upon the front bodice of clothing. Here you again divide them into two categories. One is symmetrical style lines across the bust point and the other is asymmetrical ones across the bust point.

分割线特别是功能分割线在女装上的应用较多。而且分割线的变化主要集中在前衣身的应用上。这里把它们分成两类：一类是左右对称过胸点的分割线，另一类是左右不对称过胸点的分割线。

1. Symmetrical style lines across the bust point / 左右对称过胸点的分割线

1）Princess line segmentation / 公主线分割

This type of style lines (Fig. 3-3-3) has been widely used in ladies' clothing. The lines go through the biggest change position of the convex and concave point of the body, highlighting the curve of the female body creating a perfect fit, known as the Princess line. The Princess lines on the front and back garments are two constructional seams formed by the front bust dart connecting with the waist dart and the back shoulder dart connecting with the waist dart.

这种类型的分割线（图 3-3-3）在女装中被广泛应用，因为分割线经过了人体的凹凸变化最大的部位，可以非常合身地修饰出女体的美丽线条，故它习惯被人们称为公主线。前、后片的公主缝线就是分别由前胸省和腰省、后肩省和腰省形成的前后两条结构缝线。

This style is very fitted. When copying the prototype, the full dart α needs to be considered. In detail, the bust dart α_1 (a part of the full dart α) designed to satisfy the need of the bust uplift, is transferred to be the shoulder dart, and the rest is the dart α_2 transferred to the waistline as a waist dart, and then connect the shoulder dart and waist dart with a smooth curve.

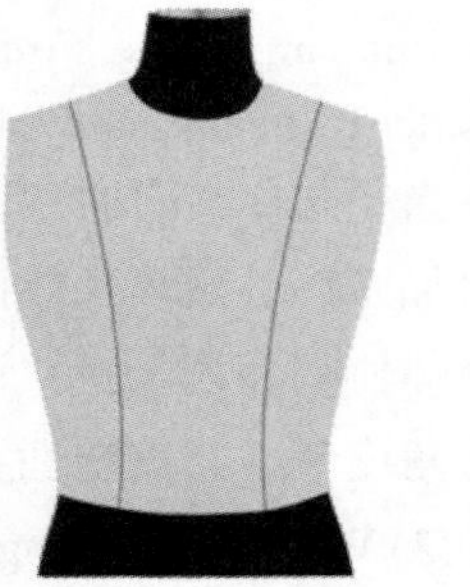

Fig. 3-3-3 The style with princess lines
图 3-3-3 有公主线分割的款式

该款式非常合体，复制原型前片基本样板时需要考虑全省 α。具体来说是将满足胸部隆起的胸省 α_1（全省中的一部分）转移为肩省，余下的 α_2 作为腰省留在腰线上，然后再将肩省和腰省以光滑的弧线进行连接。

Each step shown in Fig. 3-3-4 is as follows / 如图 3-3-4 所示步骤如下：

① Make the full dart α on the copied prototype according to the style. Transfer the bust dart α_1 to be the shoulder dart. The rest is the dart α_2 and transfer it to the waistline as a waist dart. Define the two darts needed to be connected.

根据款式图在复制的原型前片基样上做出全省 α。将其中满足胸部隆起的胸省量 α_1 转移于所需的肩部位置，余下的省量 α_2 作为腰省留在腰线上。确定被连接的两个省道。

② Connect the shoulder dart and the waist dart as a structural seam on the front piece, going through or close to the bust point. From the workmanship perspective, the shortest path connection should be adopted; while from the art perspective, the connection path needs to match the style.

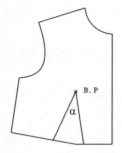 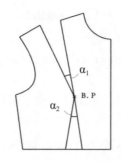 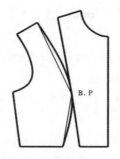 

Fig. 3–3–4 Patterns of the style with princess line
图 3–3–4 有公主线分割的款式的结构图

在前片上，将肩省和腰省连接成结构缝线，连接时尽可能通过或接近 BP 点。从工艺角度考虑，尽可能以最短路径连接；从艺术角度考虑，则要使连接线尽可能和图中的形态相符。

③ Modify the seam shape according to the style, and draw the princess lines smoothly and beautifully.

根据款式图修正缝线形态，绘制出光滑美观的公主缝线。

2）Broken line segmentation / 折线式分割

As shown in Fig. 3-3-5, it is a style with symmetrical style lines via the bust point. The style will not be particularly fitted at the waist. Therefore when copying the prototype, only the bust dart α_1 on the basic pattern is needed to satisfy the need of a bust uplift (the dart is positioned at the side seam for easy understanding), and transferred to the broken line to hide.

图 3-3-5 所示就是左右对称过胸点的分割线款式之一。该款式腰部并不特别合体，因而在复制原型前片基样时，在基样上只需做出满足胸部隆起的胸省 α_1（该省量放在侧缝处，便于理解），并将其转移到折线式分割线上隐藏。

Each step shown in Fig. 3-3-6 is as follows / 如图 3-3-6 所示步骤如下：

① Make the bust dart on the copied prototype of front piece (first transfer the bust dart to the side seams) and correct the front neck curve according to the style and draw a broken dividing line on the front piece. (The broken dividing line passes through the bust point)

Fig. 3–3–5 The style 1 with symmetrical broken line segmentation via the bust point
图 3–3–5 款式一：过胸点的对称分割线款式

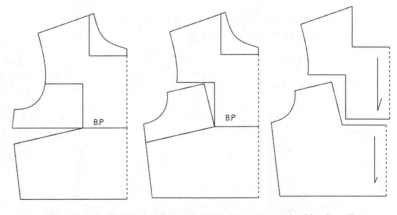

Fig. 3–3–6 Patterns of the style 1 with symmetrical broken line segmentation via the bust point
图 3–3–6 过胸点的对称分割线款式图一的结构图

在复制的原型前片上做出胸省（先把胸省量转移到侧缝线上），并根据款式图修正前领口并画出前衣片上的折线式分割线。（分割线经过ＢＰ点）

② Cut the front piece of the pattern along the broken line, fold the original bust dart, transfer the dart size to the broken line to hide.

前片沿分割线剪开基样，折叠基样上的原省道，将满足胸部隆起的胸省省量转移至分割线上隐藏。

The Fig. 3-3-7 shows the styles with symmetrical style lines across the bust points. The style will be particularly fitted at the waist. Therefore when copying the prototype, the full bust dart size is needed on the front pieces of the basic pattern (including the bust dart size and the waist dart size), and transferred to the broken line to hide.

如图 3-3-7 所示的款式中的分割线左右对称，并且通过胸点。该款式腰部特别合体，因而在复制原型前片基样时，在基样上需做出全省（包括胸省量和收腰省的量），并将其转移到分割线上隐藏。

Each step shown in Fig. 3-3-8 is as follows / 如图 3-3-8 所示步骤如下：

① Make the full bust dart (including the bust dart and waist dart) on the front of the copied prototype and draw the broken dividing line according to the style. (The broken dividing line passes the bust point).

在复制的原型前片上做出全胸省（包括胸省量和腰省量），并根据款式图画出分割线。（分割线经过ＢＰ点）

② Cut the front piece of the pattern along the broken line, fold the full bust dart, transfer the whole dart size to the broken line to hide(one part to the shoulder dart and the rest to the front center dart).

沿分割线剪开前片基样，折叠基样上的原省道，将全省量转移至分割线上隐藏（一部分转移为肩省，一部分转移为门襟省）。

③ Modify the waist seam to make it a smooth curve.
修正腰线，使之成为光滑弧线。

Fig. 3-3-7 The style 2 with symmetrical broken line segmentation via the bust point
Fig. 3-3-7 过胸点的对称分割线款式图二

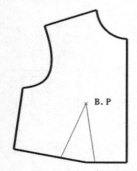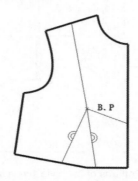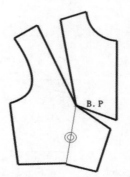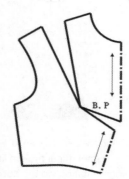

Fig. 3-3-8 Patterns of the style 2 with symmetrical broken lines via the bust point
图 3-3-8 过胸点的对称分割线款式图二的结构图

（二）Style lines not via the dart point / 不过省端点的分割线

Combination of bust dart and seam / 胸省与缝线的结合

The Fig. 3-3-9 shows a symmetrical style with no style lines through the bust points. The style will be fitted at the waist. Therefore when copying the prototype, the full bust dart value is needed on the front block of the basic pattern (including the bust dart value and the waist dart value). For a convenient constructional change, the bust dart is transferred to the shoulder dart first, and the waist dart transferred to the waist line (because the waist dart intake can be adjusted along the waistline), and then followed with other constructional changes.

如图 3-3-9 所示的款式是左右对称、分割线不过胸点的例子。该款式腰部合体，因而在复制原型前片基样时，在基样上需做出全省（包括胸省量和腰省量），为方便结构变化，先将胸省量转移到肩部，而收腰省的量就放在腰线上（因为需收的腰省可在腰线上水平移动），再进行其他的结构变化。

Each step shown in Fig. 3-3-10 is as follows / 如图 3-3-10 所示步骤如下：

① Make the full bust dart (including the bust dart and the waist dart) on the block of copied prototype, and transfer the bust dart to the shoulder, and the waist dart was left in the waist.

在复制的原型前片上做出全胸省（包括胸省和腰省），并将其中的胸省转移到肩部，腰省留在腰部。

② Make two curve-shaped style lines on the changed pattern according to the style, and put the waist dart intake on the waist line between the two style lines. Then modify the neckline curve, and make a new dart line pointing to the bust point from the style line.

根据款式图在变化后的样板上作两条弧形分割线，把腰省宽的量放入到两条分割线之间的腰围线上。然后修正领口曲线，并在分割线上作一条指向胸点的新省位线。

③ Cut the pattrn along the style line and the new dart line, close the shoulder dart on the pattern, and transfer the dart to the style line.

沿分割线和新省位线剪切，关闭样板上的肩省，使之转移至分割线上的省位线处。

Fig. 3-3-9 The symmetrical style with no style lines via the bust point
图 3-3-9 左右对称不过胸点的分割线款式

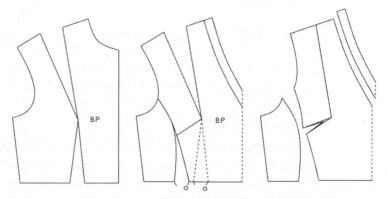

Fig. 3-3-10 Patterns of the symmetrical style with no style lines via the bust point
图 3-3-10 不过胸点的对称分割线款式的结构图

Unit 4 The changes of placket, pocket and button position on garments

第四节 服装上门襟、口袋和纽位的变化

In addition to the changes of bust dart, waist dart and style line, the front body of women's clothing also contains the changes of placket, pocket and button position. The placket, pocket and button are also important elements in clothing design, as this expresses the designer's idea.

除胸省、腰省和分割线等的变化之外，女装的前衣身还有门襟、口袋和纽位等的变化。这些局部也是服装设计中的重要元素，关系到能否恰如其分地表达设计者的设计构思。

一、Placket changing / 门襟变化

An open placket is a structure form designed on any position of garments for easy on and off access. It has a variety of forms.

服装开襟是为穿脱方便而设在衣服任何部位的一种结构形式。它有多种形式。

1. The center open placket on the front piece / 前衣片上的正中开襟

The opening in the middle of a front garment piece is full of convenience. It is simple and balanced. This is the most common site of placket opening. It can be divided into the joining placket and the symmetric overlapping placket (as shown in Fig. 3-4-1).

前衣片上的正中开襟方便，简单而平衡。这是最常见的开襟开口部位。它又可分为对合襟和叠门对称门襟（如图 3-4-1 所示）。

对合襟　　　　　单叠门对称门襟　　　　　双叠门对称门襟

Fig. 3-4-1 Several central opening styles

图 3-4-1 几种正中开襟款式

1 ）Joining plackets / 对合襟

This refers to the edge seams on the front of the left and right pattern joining with each other, without overlapping. It appears commonly in traditional Chinese-style clothing and clothing with zipper at front center, generally applicable to short jackets. Applied in Chinese-style clothing, the joining placket is usually decorated with decorative edges, and a Chinese fastener is fixed.

对合襟是指左右前片的止口合在一起，没有叠门的开襟形式。常见于传统的中式服装和前中装拉链的服装上，一般适用于短外套。对合襟应用在中式服装时，一般会在止口处配上装饰边，用中式的扣祥固定。

2）Overlapping plackets / 叠门对称门襟

It's an opening form that overlaps the two lapels. The side with buttonholes is called the placket; the other side with buttons is called the button stand. This is the most widely used form of garment placket. Generally, men's buttonholes are on the left lapel and women's are on the right.

对称门襟是有叠门的开襟形式，分左右两襟。锁扣眼的一边叫大襟或门襟，钉扣子的一边叫里襟。这是应用最广的服装门襟形式。一般男装的扣眼锁在左襟上，女装的扣眼锁在右襟上。

The width of the overlap varies with different fabric thicknesses and button sizes. The general overlap width equals the button diameter plus 0.6 cm, and the width of the button stand is 1.2–3.3 cm in general, and the buttonhole position should be on the front centerline. The overlap on a shirt made of thin fabric is generally narrow and its width is often 1.2–1.8 cm; the overlap of the Spring/Autumn clothing is made of mid-weight fabric and its width is commonly 2–2.5 cm; the thick winter clothing fabric usually has a wide overlap of 2.5–3.3 cm. The double-breasted overlap width is 5.5–10 cm in general, and buttons are usually sewed on the left and right sides of the front centerline symmetrically, but sometimes for the performance of specific modeling effect, they can be just sewed on one side.

叠门宽度因布料厚度及纽扣大小的不同而变化。一般叠门宽 = 纽扣直径 +0.6cm，单叠门宽一般为 1.2 ～ 3.3cm，其扣眼位应在前中线处。一般薄型面料制作的衬衫类服装叠门较窄，常取 1.2 ～ 1.8cm；中厚型面料制作的春秋装类的叠门以 2 ～ 2.5cm 为常见；厚型面料制作的冬装则取较宽的叠门量，即 2.5 ～ 3.3cm。双叠门的宽度一般为 5.5 ～ 10cm，纽扣一般对称地分列在前中线左右两侧，但有时为表现特定的造型效果，也会仅钉在一侧。

The overlapping often presents two kinds of forms: placket front and concealed placket. When buttons can be seen from the front, it is called the placket front. If the buttons can not be seen from the front, and are sewn to the under placket layer, that is called a concealed placket.

单叠门又有明门襟和暗门襟之分。正面能看到纽扣的称为明门襟。正面看不到纽扣且纽扣被缝在衣片夹层上的称为暗门襟。

2. Open placket forms in other parts / 在其他部位的开襟形式

The open placket takes many forms. In addition to being on the garment front, the open placket can also be on the back, shoulder, underarm etc. (Fig. 3-4-2, Fig. 3-4-3)

衣服的开襟形式有很多。除了在衣服正面开襟的形式外，还可在后面开襟、在肩部开襟和在腋下开襟等。（图 3-4-2、图 3-4-3）

Fig. 3-4-2 The open placket on the
shoulder and underarm
图 3-4-2 肩部和腋下开襟款式

Fig. 3-4-3 The open placket on the back
centerline
图 3-4-3 后中开襟款式

3. Placket shape changing / 门襟造型变化

The placket shape can be changed in many ways. Besides symmetric lapel and asymmetric lapel forms (as shown in Fig.3-4-4), it can also be divided into the linear lapel, slash placket and curve lapel according to its shape. Furthermore, it can be divided to half and full placket forms according to its length. The slash placket and curve placket are used more in "QiPao" and half plackets are used more in knitted garments, shirts and pullovers.

门襟的造型变化有多种，除了可分为对称襟和非对称襟外（如图 3-4-4 所示），按其形态还可分为直线襟、斜线襟和曲线襟等。按其长短还可分为半开襟和全开襟等形式。斜线襟和曲线襟在旗袍上应用比较多，半开襟在针织服装、T- 恤衫和套头衫上应用比较多。

When designing structures for asymmetric placket patterns, careful observation should be made to see the style so as to be expressed accurately on the base sample.

在对非对称门襟款式进行结构设计时，要仔细观察款式图，以便准确地在基样上表达出来。

Fig. 3-4-4　Changing chart for asymmetric placket shaping
图 3-4-4 非对称门襟造型变化图

二、Pocket changing / 口袋变化

The pocket is one of the main garment accessories. Its main function is as a hand device, and to containing artifacts as well as having a role as a style feature.

口袋是服装的主要附件之一，其功能主要是放手和装盛物品，以及起点缀装饰美化的作用。

1. Pocket classification/ 口袋分类

The pocket is a general term, and used widely in clothing, and each pocket name is different. But for the structure, it can be classified into three types.

口袋是一个总称，在服装上的应用很多，名称各异，但从结构上来分可归纳为三大类。

1）Insert pocket/ 挖袋或嵌线袋

The insert pocket is a structure from where a piece is cut out of the clothing according to the pocket size, and pocket fabric is sewn inside, also known as an opening pocket. As shown in Fig. 3-4-5, according to the sewing technology of the pocket opening, the insert pocket can be divided into several forms: single block line, double inlay line, box pocket, etc. Some are still decorated with various pocket covers. According to the shape of the pocket opening, the insert pocket can be divided into straight, transverse, inclined type, curve form and so on. If it is very difficult for the opening of the curve, the form should be smooth and docile in the production process. It is rare in practical production. The insert pocket is often used to categorize clothing, such as dresses, suits, student uniforms, casual clothes and so on.

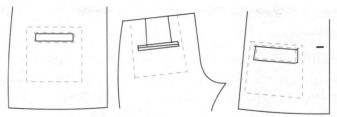

Fig. 3–4–5 Several insert pocket shapes
图 3–4–5 几种挖袋造型图

挖袋是一种在衣片上剪出袋口尺寸并内缝袋布的结构形式，又称开袋。如图 3-4-5 所示，从袋口的缝制工艺来分有单嵌线袋、双嵌线袋、箱型口袋等，有的还装饰各种各样的袋盖；从袋口形状分有直列式、横列式、斜列式、弧形式等，但弧形式袋口在工艺制作上很难做得平整伏贴，在实际应用上比较少见。挖袋常用于外套类服装上，如礼服、西服、学生服以及便装等。

2) Hidden pockets / 插袋

The hidden pocket generally refers to a hidden pocket in the seam, and usually it is unnecessary to cut the fabric. This kind of pocket is good for concealment. It may also be added to the stitch line, with a pocket flap covering the edge, such as the pocket on a ladies' princess line or knife-back line, and the side bag of male trousers etc. In addition, there are slant pockets in male and female trousers; slant pockets are more convenient than straight pockets.

插袋一般是指在服装分割线缝中留出的口袋，一般不用剪开衣片。这类口袋隐蔽性好，也可缉明线，加袋盖或镶边。如女装公主缝线或刀背缝线上的插袋，男西裤上的侧袋等，另外，男、女西裤上还有斜插袋，斜插袋比直插袋更方便插手。

3) Patch pockets / 贴袋

The patch pocket is a pocket made of fabric sewed onto the surface of the garment (as shown in Fig. 3-4-6). According to its structure, it can be divided into cover, no cover, letter patch pocket and open pocket (another insert patch will be made in the patch) etc. From this process, it can be divided into two forms: decorative stitch and no decorative stitch. Shaping can be changed dramatically, for example, taper-angle shapes, rounded corner shapes, circles, ovals, rings, crescents, and various other irregular shapes. Patch shapes also include invisible pleat and visible pleat garment bags.

贴袋是用面料缝贴在服装表面上的一种口袋（如图 3-4-6 所示）。在结构上大致可分为有盖、无盖，子母贴袋和开贴袋（在贴袋上再做一个挖贴袋）等；在工艺上可分为缉装饰缝和不缉装饰缝两种；造型上则可千变万化，可做成尖角形、圆角形、圆形、

Fig. 3-4-6 Several patch pocket shapes
图 3-4-6 几种贴袋造型图

椭圆形、环形、月牙形及其他各种不规则形等。贴袋造型还包括暗裥袋、明裥袋等。

2. Pocket design / 口袋设计

Pockets are so widely used in clothing because of the double properties: It is functional and decorative. But while designing, some consideration should be given to the following:

衣袋以其在服装上的双重特性：功能性和装饰性，因而使用广泛。在进行衣袋的设计时应考虑如下：

(1) The design of the pocket opening size should be based on the size of the hand. The general width of an adult female hand is 9–11 cm, and the width of an adult male hand is 10–12 cm. The net dimension for men's and women's jacket pocket bags is determined according to the hand width plus roughly 3 cm. In addition, considering the process requirements, if top-stitching, you should also add the line width. The volume of the pocket opening can also be increased for the straight pocket of a coat and pants.

衣袋的袋口尺寸应依据手的尺寸来设计，一般成年女性的手宽在：9 ～ 11cm 之间；成年男性的手宽在：10 ～ 12cm 之间。男女上衣大袋袋口的净尺寸一般可按手宽加放 3 cm 左右来确定。另外，再考虑工艺上的要求，如果缉明线，还应加明线的宽度。对大衣类服装和裤子的直插袋，袋口的加放量还可增大些。

(2) The design of the pocket position should be generally coordinated with the overall style of the clothing, taking into account the whole garment balance. Generally, the opening height of the big pocket takes the bottom line for reference, the length measuring 1/3 upward, minus 1.3–1.5 cm, or at the position below the waist line about 5–8 cm. Because of the longer length, and according to the style required, the coat pocket position can also be moved down to a position below the waist line 9–10 cm. The front and rear parts of the pocket opening should be set forward upon the front chest width line 1.5–2.7 cm as the center. The height of the small chest pocket of the jacket and the front end of the chest pocket of the Chinese tunic suit should aim at the second button. The front end of the chest pocket on the suit reference to the bust line up about 1–2 cm, and the back end of the pocket is 2–4 cm away from the chest width line.

袋位的设计一般应与服装的整体造型相协调，要考虑到使整件服装保持平衡。一般上装大袋的袋口高低以底边线为基准，向上量取衣长的三分之一减去 1.3 ～ 1.5cm 或在腰围线下 5 ～ 8cm 左右位置。但大衣因其衣长较长，根据款式需要袋位还可适当下降，可定在腰围线下 9 ～ 10cm 位置。袋口的前后位置以前胸宽线向前 1.5 ～ 2.7cm 为中心来定。上衣胸袋的袋口高低：中山装胸袋袋口前端对准第二粒纽位，西服上胸袋袋口前端参考胸围线向上 1 ～ 2cm 左右，胸袋口的后端距胸宽线 2 ～ 4cm。

(3) The other characteristics of the pocket itself should be taken into account. The pocket shape should especially be proportional to the garment shape, but also change with the specific requirements of certain styles. In the conventional design, generally the bottom of the pocket is slightly larger than the pocket opening, and the pocket depth should be slightly larger than the bottom of the pocket. In addition, the pocket material, color, patterns should also be coordinated with the garment, so as to achieve the ideal decorative effect.

要考虑衣袋本身的造型特点。特别是贴袋的外形，原则上要与服装的外形成正比，但也要随某些款式的特定要求而变化。在常规设计中，一般贴袋的袋底稍大于袋口，而袋深又稍大于袋底。另外，贴袋的材质、颜色、花纹图案还应与整个服装相协调，才能达到较理想的装饰效果。

三、Change of button position / 纽位变化

The placket change determines the button position. The arrangement of button at placket is usually equal, but for special long garments, the spacing of the button position should get bigger the lower down, otherwise the intervals will not seem equal.

门襟的变化决定了纽位的变化。纽位在叠门处的排列通常是等分的，但对衣长特别长的衣服，其间距应是愈往下愈长。否则其间隔看来是不相等的。

For the general jacket, the key is to determine the position of the top and bottom buttons. The top button position is related to the style of the garment. For the determination of the bottom button position, different types of clothing have different references: Shirts are often based on the bottom line, and take 1/3 of the length up minus about 4.5 cm; For suits or jackets, the bottom button position is usually flush with the pocket opening line.

对一般上装而言，最关键的是最上和最下一粒纽位的确定。最上面一粒纽位与衣服的款式有关。最下一粒纽位的确定，不同种类的服装确定其纽位的参照不同：衬衫类，常以底边线为基准，向上量取衣长的三分之一减 4.5cm 左右来定；套装或外套类服装，常与袋口线平齐。

Buttons can be divided into two forms: functional buttons and decorative buttons. Functional buttons are used to fasten garments and pockets, both functional and decorative. Decorative buttons are purely decorative buttons sewn on the chest, pockets, collars, sleeves, etc. Buttons are generally arranged one by one, and can also be arranged in groups of 2 to 3 buttons. The button midpoint is generally on the front centerline of the garment.

纽扣按其功能可分为功能扣和装饰扣两种。功能扣是指扣住服装开襟、衣袋等处的纽位，兼有实用性和装饰性功能；装饰扣是指缝钉在前胸、口袋、领角、袖子等部位的纯粹起装饰作用的纽扣。纽扣一般一粒一粒单个排列，也可 2～3 粒一组一组排列。纽扣的中点一般在衣服的前中线上。

The buttonhole position is not completely the same. For example, a woman's shirt with a man's collar; its placket has a reverse structure. The buttonholes are all vertical except one that is horizontal at the collar. The vertical buttonholes are on the mid-line; the front end of horizontal buttonhole is 0.2–0.3 cm away from the mid-line. For other garments such as suits, the buttonhole is generally horizontal, the front end of the horizontal buttonhole is generally 0.3–0.4 cm away from the mid-line. (This changes as the fabric changes in thickness and size and the width of the button). As shown in Fig. 3-4-7.

纽眼的位置并不完全与纽扣相同，如男式衬衫领女衬衫，其门襟是外翻边的结构，其纽眼位除了在领上的一颗

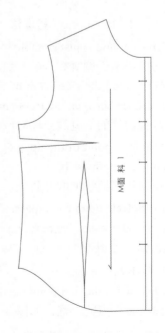

Fig. 3–4–7 Buttonhole position chart
图 3-4-7 纽眼的位置图

是横向外，其余的都是纵向，纵向的纽眼位在前中线的位置上，横向的纽眼前端偏出中线 0.2 ～ 0.3cm；而其他的衣服或一些外套类服装如西服，其纽眼一般是横向的，横向的纽眼前端一般偏出中线 0.3 ～ 0.4cm（视面料的厚薄和纽扣的大小厚度而变化）。如图 3-4-7 所示。

Unit 5　Developing bodice patterns with CAD software
第五节 用 CAD 软件制作衣身变化纸样

In this section, the methods of dart transferred, pleated structure change and style line change will be introduced with CAD software, by using Japanese Bunka Prototype for Women. You can copy some prototypes of the front and back bodices in advance.

本节依托文化式女装的衣身，介绍用 CAD 软件进行省道转移、褶裥结构变化和分割线结构变化的方法。事先可复制出若干个前、后片衣身原型备用。

一、Dart transferred with CAD software / 省道转移 CAD 应用

There are three methods of dart transferred: measuring, rotating and cutting. The cutting way is the same as the rotating. Here, the ways of measuring and rotating will be introduced as follows:

省道转移的方法有三种：量取法、旋转法以及剪开法，其中旋转法与剪开法类似。下面只介绍旋转法的操作方法。

（一）Measuring / 量取法

This method is just suitable for the dart that will be made in the side seam.

它仅适用于侧缝线上的省道。

(1) Extend the side seam of the front piece to the extention line of the front waist line. Measure the difference between the front and back side seams as "a". As shown in Fig.3-5-1.

延长前片的侧缝线至前腰围的延长线。测量前后侧缝线之差值 a。如图 3-5-1 所示。

(2) Get a point of the opening dart on the front piece. Join it and the bust point as a dart line and measure its length as "b".

在前片侧缝线上取开省点，把它与 BP 点连接成开省线，测量其省边长为 b。

(3) Select "double compasses" – input "a, b" – click in the blank space, and you can get the side seam dart. Draw the side seam line of the lower segment. As shown in Fig. 3-5-2.

选择"双圆规"—输入"a，b"—单击空白处，即得侧缝省。画出下段侧缝线，如图 3-5-2 所示。

(4) Correct the dart point position. As shown in Fig. 3-5-3.

修正省尖点的位置。如图 3-5-3 所示。

（二）Rotating / 旋转法

First you must determine the bust dart angle α_1, then introduce the way of transferring dart by

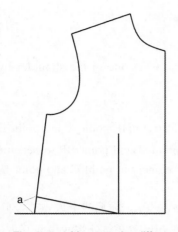

Fig. 3–5–1 Measure the difference
between the front and back side seams
图 3–5–1 测量前后侧缝差值

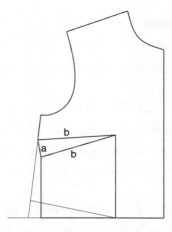

Fig. 3–5–2 Make the side seam dart
图 3–5–2 作侧缝省

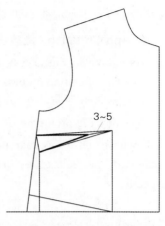

Fig. 3–5–3 Correct the dart
point
图 3–5–3 修正省尖

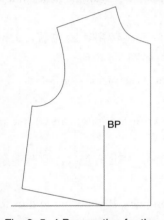

Fig. 3–5–4 Preparation for the
prototype of front piece
图 3–5–4 原型前片准备

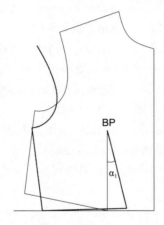

Fig. 3–5–5 The angle formed by rotating
图 3–5–5 旋转角度

rotating.

首先定出胸省量 α_1，再介绍旋转法转移省道。

1. Measure α_1 / 测量 α_1

(1) Take a copied front prototype. Draw a vertical line from the bust point to the waist line. Extend the waist line. As shown in Fig. 3-5-4.

选用一个复制好的前片原型，从胸点向下至腰围线作垂线，延长腰围线，如图 3-5-4 所示。

(2) In an empty state, select the armhole curve, side seam and waist line, and rotate them around the bust point until the side seam contacting the waist line. The angle formed by rotating is displayed in the data bar and it is the bust dart angle α_1. Draw another leg of the bust dart α_1 using the tool of "Angle line", as shown in Fig. 3-5-5.

空状态下，选择前片的袖窿弧线、侧缝线、腰围斜线等，以 BP 点为中心旋转直至侧缝线触碰到腰围线。此时数据框内显示的旋转角度，即为胸省量 α_1 的值。用"角度线"工具画出

胸省 α_1 的另一条边线，如图 3-5-5 所示。

2. Apply rotating / 旋转法应用

As shown in Fig. 3-5-6, a dart in the waist can be transferred to the one in the shoulder by rotating. The operating process is as follows：

如图 3-5-6 所示，在腰部的省道可利用旋转将其转移到肩部。操作过程如下：

① Copy the prototype of the front body first, make the bust dart α_1, mark point A and point B as shown in Fig. 3-5-6; take point C on the shoulder line; connect point C and point BP to become a new dart line. Select the "Break off a Line" tool to break the line segment at point C and point B, respectively.

先复制前衣身原型，作出胸省 α_1，按图 3-5-6 所示标出点 A、点 B；在肩线上取点 C；连接 C 点和 BP 点成为一条新省线。选择"断开线段"工具，分别在 C 点和 B 点处将线段断开。

② Select the lines from point B to C clockwise. Rotate from point B to A with the bust point as the center. At the same time, point C will be rotated to point C'. Connect point C' with the bust point to become another leg of the new dart. As shown in Fig. 3-5-7.

按顺时针方向选择从 B 到 C 点之间的线段，以 BP 点为中心将 B 旋转到 A 点位置，此时 C 点被转移到 C' 点，连接 C' 点与 BP 点即为新省的另一条边线。得如图 3-5-7 所示肩省。

③ Modify the waist line and dart point. Finishing of transferring dart is shown in Fig. 3-5-8.

修正腰围线和省尖。省道转移完成如图 3-5-8 所示。

The angle of the dart won't change, but the width will change with the length in course of transferring. Then modify the dart and delete useless lines .

省道在转移过程中角度始终保持不变，省宽随省长的变化而变化。修正省道线，删除多余线段。

Note: these two ways of transferring dart are limited to construction lines. If you want to transfer a dart on the pattern, it is necessary to make a dart by using Zhizunpen. Then select "dart" in "handle pattern" – "rate rotating dart".

注意：这二种省道转移限于结构线。如果要在样片上操作则需要用"智尊笔"挖好省，然后再用"样片处理"中的"省"—"比率转省"。

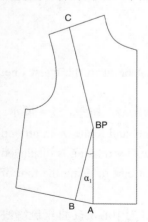

Fig. 3-5-6 Bust dart in waist
图 3-5-6 胸省在腰部

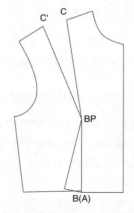

Fig. 3-5-7 Transfer dart to shoulder
图 3-5-7 胸省转移至肩部

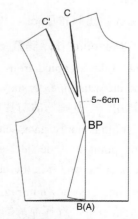

Fig. 3-5-8 Modify dart and waist line
图 3-5-8 修正肩省及腰线

（三）Example for dart transferred / 省道转移实例

1. Transfer bust dart to armhole / 胸省转移到袖窿

As shown in Fig. 3-5-9(1), this style is loose at the waist. It only considers the bust dart α_1 and transfers it to the armhole.

如图 3-5-9(1) 所示，该款式腰部宽松，只需考虑将胸省量 α_1 转移到袖窿部位即可。

The process is as shown in Fig. 3-5-9(2) / 过程如图 3-5-9(2) 所示：

① Copy the prototype of the front body first, make the bust dart α_1, mark point A and point B as shown in Fig. 3-5-9(2).

先复制前衣身原型，作出胸省 α_1，按图 3-5-9(2) 所示标出点 A、点 B。

② Make a point C on the armhole curve. Connect point C and the bust point, and select "Break off a line" tool to break the armhole curve at point C.

在袖窿弧线取点 C，与 BP 点连接，选择"断开线段"工具，在 C 点断开袖窿弧线。

③ Select the lines from point B to point C clockwise. Rotate them with the bust point as the center to move point B to point A . At the same time, point C will be rotated to point C'. Connect point C' and the bust point as the other leg of the new dart.

按顺时针方向选择从 B 点到 C 点之间的线，以 BP 点为中心进行旋转，使 B 点转移到 A 点位置。此时 C 点被转到 C'。连接 C' 与 BP 点即成为新省的另一条边线。

④ Modify the waist line and the dart point.

修正腰围线及省尖位置。

(1) the style/ 款式

 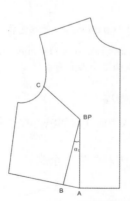  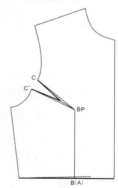

(2) the process / 转移过程

Fig. 3-5-9 The process of transferring bust dart to armhole

图 3-5-9 胸省转移至袖窿

2. Transfer one bust dart into two or more / 胸省量转二个省或多个省

As shown in Fig. 3-5-10(1), it is an example that the full bust dart will be transferred into two curve darts in the shoulder and one waist dart. It needs transferring on the pattern.

如图 3-5-10(1) 所示，这是将一个全胸省转移成移到肩部的二个弧形省和腰省的应用。由一个省转移成多个省需要在样片上操作。

The steps shown in Fig. 3-5-10(2) are as follows / 如图 3-5-10（2）所示步骤如下：

① Copy the prototype of the bust dart on the front piece, and take it out into the pattern. Select the "Dart" tool – "Define Dart" and define the two angle lines as dart lines.

复制含胸省量的前片结构图，转化成样片。选"省道"工具—"定义省"，将 2 条夹角线定义为省道线。

② Draw two curves on shoulder line.

在肩线上按要求作两条弧线。

③ Right-click dart line in an empty state – "ratio transfer dart" – select the center and old dart line, right-click to end – select the new dart line anti-clockwise – input "1/2" – enter.

空状态下右击省道线—"比率转省"—按提示选择省中心、旧省线—右击结束—按逆时针方向选择新省线—输入比率"1/2"—回车。

④ Adjust the pattern to be vertical.

调整样片至垂直方向。

⑤ Dig a dart on the waist line. Its width is according to the measurement.

在腰围线上挖腰省量。腰省宽可根据实际计算而得。

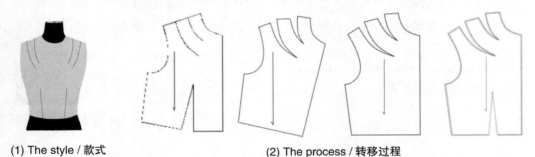

(1) The style / 款式   (2) The process / 转移过程

Fig. 3-5-10 Transfer bust dart into two shoulder curve dart
图 3-5-10 胸省转移为二个弧形肩省

3. Transfer of the full bust dart / 全胸省量的转移

As shown in Fig. 3-5-11(1), the style is a fitted waist. The structure needs to include the full bust dart.

图 3-5-11（1）所示的款式腰部很合体，其结构需要包含全胸省。

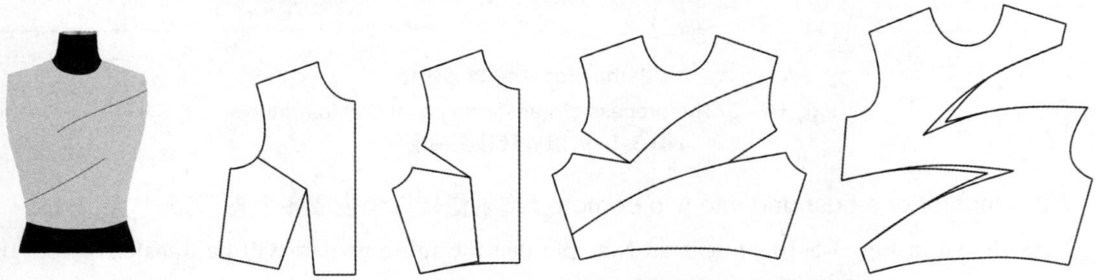

(1) The style / 款式   (2) The processs / 转移过程

Fig. 3-5-11 Example 1 of the full bust dart transferred to two asymmetric darts
图 3-5-11 非对称全胸省转移款式一

At first, the full bust dart can be transferred to a temporary position. Then transfer it to the required position. As the two darts are asymmetric, you need to unfold the whole front piece symmetrically first.

首先将全胸省转移至临时省位，再转移到要求位置。因省道不对称，要先对称展开前片。

The process shown in Fig. 3-5-11(2) is as follows/ 如图 3-5-11(2) 所示操作过程如下：

① Copy the prototype of the front piece with the full bust dart on it. Take a point on the armhole curve, and connect it and the bust point. Cut the armhole curve at this point.

复制含全胸省量的原型前片结构图，取袖窿弧线上一点并与 BP 点连接，并在此点处切断袖窿弧线。

② Transfer the full bust dart to the armhole temporarily. Draw the waist line smoothly. Take the changed structure out into the pattern.

将全胸省暂先转移到袖窿。修顺腰围线。提取变化的结构图，形成为样片。

③ Use the "Symmetrical Unfold" tool to unroll the whole front piece, then cancel the symmetry of the sample piece, and make two new dart lines pointing to the bust point according to the style.

用"对称展开"工具展开整个前片，然后取消样片对称，根据款式作出两条指向 BP 点的新省位线。

④ Transfer the whole bust dart to the new dart lines. Modify the dart points and the dart legs.

将省道转移至新省位线中。修正省尖点的位置及修顺省的两条边线。

Fig. 3-5-12(1) is an example that the full bust dart will be transferred asymmetrically too. Because the new dart line connects with the old one, transfer one dart only at a time. The process is as shown in Fig. 3-5-12(2).

图 3-5-12(1) 所示的款式也属于不对称的全胸省量转移的实例。由于新省线与旧省线相连，只需要一次转移就能完成。图 3-5-12(2) 所示为该款式转移过程图。

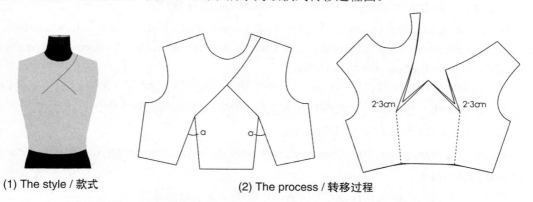

(1) The style / 款式　　　(2) The process / 转移过程

Fig. 3-5-12　Example 2 of the full dart transferred to two asymmetric darts
图 3-5-12 非对称全胸省转移款式二

二、Design and change of a pleat / 褶裥设计及变化

（一）Gathers / 碎褶

Fig. 3-5-13(1) is a style with gathers combined with waist darts, created a fitted waist. The structure needs to include the full bust dart.

如图 3-5-13（1）所示是碎褶与腰省相结合的款式，且腰部非常合体，因而其结构需要包含全胸省。

The steps are as follows / 具体步骤如下：

① Copy the prototype of the front piece with the full bust dart and take it into the pattern.

复制含全胸省量的原型前片并取成样片。

② Draw the auxiliary lines on the dart line .

在省边线上作辅助线。

③ Select tool of "cut and move" in "handle pattern" and cut along the auxiliary line to get a result as shown in Fig. 3-5-13(2). Modify the dart point and the dart lines.

选择"样片处理"工具中的"剪开移动"沿辅助线剪开，得图 3-5-13（2）所示结果，修正省尖点的位置及修顺省的两条边线。

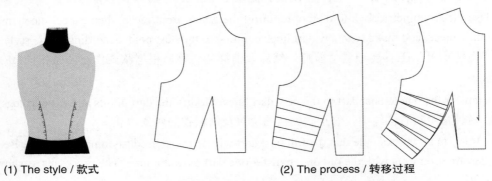

(1) The style / 款式　　　　　　　　(2) The process / 转移过程

Fig. 3-5-13 Patterns of gathers on waist dart

图 3-5-13 腰省上的碎褶结构图

（二）Apply and change the pleats / 褶裥的应用及变化

The way to make pleats is the same as making gathers. Fig. 3-5-14(1) is a style where there are three vertical pleats on the front pattern. There are curves cutting in the shoulders and loose at the waist. Only transfer the bust dart to three pleats.

褶裥的制作方法与碎褶类似。如图 3-5-14（1）所示的款式在前衣身上设计了三个纵向褶裥。肩部有弧形分割线，腰部宽松，只需将胸省量转移至三个纵向的褶裥当中。

The steps are as follows / 具体步骤如下：

① Copy the prototype of the front piece with only the bust dart on it. Draw a line from the bust point to the side seam.

复制只含胸省量的原型前片结构图。从 BP 点画一条线到侧缝线。

② Transfer the bust dart to the side seam by rotating.

用旋转法将胸省转移到侧缝。

③ Draw the cutting line and cut the front into two parts. Take out the two patterns.Use the tool "measure length of curve" to measure the cutting line, and mark as ＊ . Draw an auxiliary line from the bust point to the cutting line on the lower piece.

作分割线将前片分为上下二部分，并分别取出成为样片。测量分割线长度，标记为＊。在下片上从 BP 点画一条辅助线到分割线处。

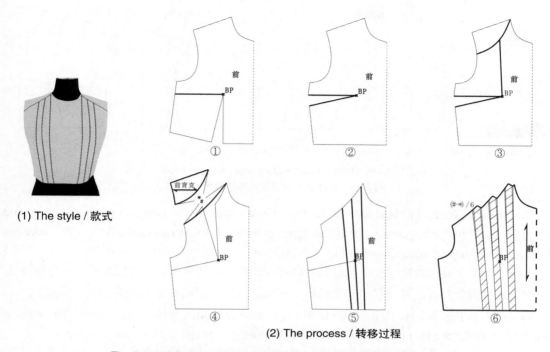

(1) The style / 款式

(2) The process / 转移过程

Fig. 3–5–14 Patterns of the front piece with three longitudinal pleats

图 3–5–14 前衣身有 3 个纵向褶裥的结构图

④ Define the side seam dart. Transfer the side seam dart to the auxiliary line using "ratio transfer dart" and adjust the curve. Use the tool "measure length of curve" to measure the curve, and mark as ＃.

将侧缝省"定义省"，利用"比率转省"将侧缝省转移到辅助线处，并修正弧线。用"测量线段长度"工具测量弧线长度，标记为＃。

⑤ Copy the changed front piece , and draw three auxiliary lines from the curve to the waist line on its structure drawing , that is, the pleat position. The upper and lower spacing of the three auxiliary lines is not equal.

复制已变化的前片结构图，并在结构图上从弧线作 3 条辅助线至腰围线，即为褶裥位。这 3 条辅助线的上、下间距不等。

⑥ Select "handle pattern" – "pleat" – "making pleat", and add the amount of pleats required. Finally, the amount of "(# - *) /3" is averaged to be added to each side of the upper end of the 3 pleats (above the bust point). The making process is shown in Fig. 3-5-14(2).

选择"样片处理"—"褶"—"褶生成"工具，加放出所需的褶裥量。最后，将"（ ＃ － ＊ ）/3"的量平均加到 3 个褶裥的上端的两侧（BP 点以上）。图 3-5-14（2）所示为制作过程。

三、Design and developing of styling line / 分割线设计及变化

It is easy to cut patterns in CAD software. The shape of the styling line is important. The Princess line is typical, as shown in Fig. 3-5-15(1).

在 CAD 软件中分割操作比较简单，关键在于样片中分割线形状的设计。典型分割线有公主线分割，如图 3-5-15(1)。

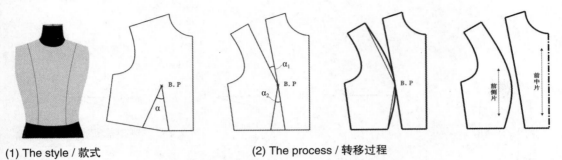

(1) The style / 款式　　　　　　　　(2) The process / 转移过程

Fig. 3-5-15 Patterns of the style with princess lines
图 3-5-15 有公主线的款式的结构图

The steps: make the full bust dart on the front piece – transfer the bust dart part to shoulder and don't move the waist dart – connect both dart lines as styling lines and make them smooth – take out each piece into the pattern, as shown in Fig. 3-5-15(2).

操作步骤：在前片结构图模板中做出全胸省—胸省转移到肩部，腰省位置不变—分别连接两个省道线作为分割线，并使它们光滑圆顺—分别取出样片即可，如图 3-5-15（2）所示。

As shown in Fig. 3-5-16(1), it is a style that is symmetric with a broken cutting line. The waist is loose. Only transfer the bust dart to the broken cutting line. The shape of the collar is square.

如图 3-5-16（1）是左右对称的分割线款式。该款式腰部并不特别合体，因而只需要将胸省量转移到分割线上隐藏。领形为方领。

The step are as follows / 操作步骤如下：

① Make a square collar and draw a broken cutting line and transfer the bust dart to the broken cutting line.

作方领和分割线，并将胸省转移到分割线。

② Along the broken cutting line, take the upper and lower parts into pattern to make them into upper and lower pieces, as shown in Fig. 3-5-16(2).

沿着分割线，分别将上、下两部分取样片，使之成为上、下片，如图 3-5-16（2）所示。

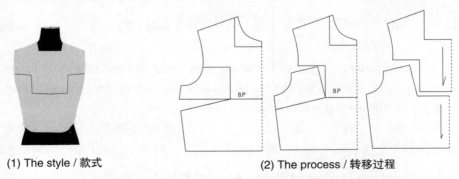

(1) The style / 款式　　　　　　　　(2) The process / 转移过程

Fig. 3-5-16 Patterna for the style 1 with symmetrical broken line via the bust point
图 3-5-16 过胸点的对称分割线款式的结构图

Exercise/ 练习

1. Make patterns of front piece according the pictures as follows (Dart transfer).

根据下面的款式图做出前片样板（省道转移）。

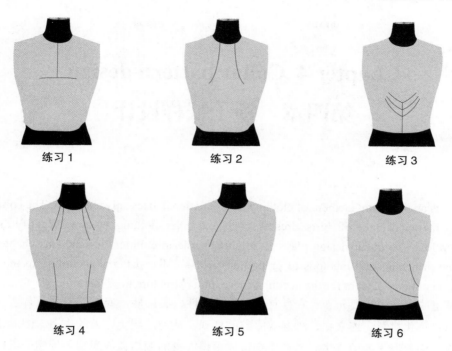

练习 1 练习 2 练习 3

练习 4 练习 5 练习 6

2.Make patterns according the pictures as follows (Application of pleats) .

根据下面的款式图做出前片或后片样板（褶裥变化）。

练习 7 练习 8 练习 9

3.Make patterns according the pictures as follows (Application of style lines) .

根据下面的款式图做出前片样板（分割线变化）。

练习 10 练习 11 练习 12

Chapter 4　Collar pattern design
第四章　领子纸样设计

A collar is an important element of clothing. It is functional, decorative, or both. The collar design can beautify the appearance of shirts, dresses, coats and other clothing, so as to beautify the role of people. Therefore, the collar design plays an important role in clothing design. This chapter mainly describes the principles and methods of preparing various collar patterns according to the neckline curve of garment body. Except for the neckline curve, other structure lines are omitted.

衣领是组成服装的重要元素。它具有一定的功能性、装饰性，或者二者兼而有之。通过对领子设计可以美化衬衫、连衣裙、外套等服装的外观，从而达到美化人的作用。因此，领子设计在服装设计中起着重要的作用。本章主要讲述根据衣身的领口弧线配制各种领子纸样的原理和方法，除了领口弧线外衣身其他结构都省略。

Unit 1　Basic knowledge of collar pattern design
第一节　领子纸样设计基础知识

一、Types of collars / 领子的种类

There are three basic types of collars: the collarless, stand collar, fold-over collar. The three basic collars can be combined with the garment body and pleats and darts to form a variety of developing collars.

领子有三种基本类型：无领、立领、翻折领。三种基本领型与衣身、褶裥、省道结合可以构成各种变化领型。

1. Collarless / 无领

"Collarless" can also be called the "neckline collar" as it is the neckline shape. According to the structure, it has two types: the front opening and the crossing head. As shown in Fig. 4-1-1, common styles of Collarless have the round neckline, boat neckline, V-shaped neckline, square neckline, etc.

"无领"亦称"领口领"，即无领身部分，且以领口的形状为衣领造型线。根据构造它有前开口型和贯头型两种。如图 4-1-1 所示，常见的无领有圆领、船形领、V 形领、方形领等。

2. Stand collars / 立领

Stand collars are divided into single stand collars and turndown stand collars. The turndown stand

Fig. 4–1–1 Examples of collarless design
图 4–1–1 无领款式图

collar consists of the part that stands up inside and the part that turns down outside.

立领分为单立领和翻立领。翻立领主要包括里面立起来的部分和外面翻下来的部分。

3. Fold–over collars / 翻折领

The fold-over collar consists of a collar band and a fold (or a fall). There are two types of fold-over collars. One is an ordinary fold-over collar, whose collar band is connected with the fold, such as the collars on various jackets and coats; Another type is that the collar band is separated from the fold (or the fall), as is often the case on shirts and suits. The fold on the suit and the lapel on the garment body join together, and that is called lapel collar.

翻折领则由领座和翻领部分组成。翻折领分为两种类型。一种是普通翻折领，其领座部分与翻领部分连在一起，如各种夹克和外套上常用的领型；另一种是领座部分与翻领部分分开的类型，如衬衫上和西装外套上常用的领型。西装外套上的翻领和衣身上的驳头合在一起，被称为翻驳领。

4. Developing collars / 变化领

The three basic collar types can be combined with pleats and darts to form a variety of developing collars. Developing collars are commonly used in fashion design. Such as cowl collar and frilled collar belong to the developing collars that comes from the base collar type change.

三种基本领型与褶裥、省道结合可以构成各种变化领型。变化领常用在时装设计上。如垂褶领、波浪领都属于由基础领型变化而来的变化领。

二、Common names on collar / 领子上常用名称

Different collar types have different structures and use different names. The common names in collar pattern design are Neckline, Assembling Neckline, Collar Band, Lapel, Fold Line or Roll Line, Style Line or Shape Line, Break Point, Cocking up etc. The names that are often used to draw the lapel collar are shown in Fig. 4-1-2.

各种类型领子的结构不同，用到的专有名称也不同。领子纸样设计中常用的名称有领口线、装领线、领座、翻领、翻折线、款式线或轮廓线、驳点、起翘等。如图 4-1-2 所示为绘制翻驳领时常用到的名称。

1. Neckline / 领口线

The line where the collar is stitched to the garment body. It is on the garment body.
领子和衣身上领口缝合的线。领口线在衣身上。

2. Assembling neckline / 装领线

The line where the collar is stitched to the garment body. It is on the collar.
领子和衣身上领口缝合的线。装领线在领子上。

3. Collar band / 领座

The stand-up part from the neckline to the roll line.

由领口线至翻折线的直立部分。

4. Fall / 翻领

The part from the roll line to the collar out edge.

由翻折线至外轮廓线之间的部分。

5. Roll Line / 翻折线

The line where the collar rolls over.

领子翻折的线。

6. Style Line / 款式线或造型线

The outer edge of the collar or rever.

领子或者翻领的外轮廓线。

7. Break Point / 驳点

The point where the rever turns back to form a lapel.

驳头翻折的最低位置。

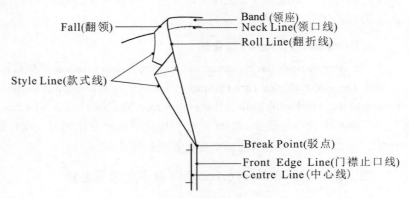

Fig. 4-1-2 Names on the lapel collar

图 4-1-2 翻驳领的专有名词

Unit 2 Collarless Design

第二节 无领设计

一、Collarless / 无领

1. A round neckline / 圆领

On the basis of the base neckline, increase the front neckline depth by 6.5 cm and also increase the front neckline width by 4 cm. Increase the back neckline depth by 3 cm and also increase the back neckline width by 4 cm. Draw the new neckline shape with a smooth curve, as shown in Fig. 4-2-1.

　　如图 4-2-1 所示，在基础领口基础上，将前横开领开大 4cm，前直开领开大 6.5cm，将后横开领开大 4cm，后直开领开大 3cm。画顺新领口线。

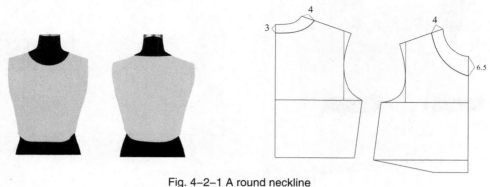

Fig. 4-2-1 A round neckline
图 4-2-1 圆领结构制图

2. A boat neckline / 船形领口

As shown in Fig. 4-2-2, on the basis of the base neckline, reduce the front neckline depth by 1cm and increase the front neckline width by 6cm. Increase the back neckline depth by 3 cm and the back neckline width by 6 cm. Draw the new neckline shape with a smooth curved line.

　　如图 4-2-2 所示，在基础领口线的基础上，将前直开领抬高 1cm，前横开领开大 6cm；将后领横开领开大 6cm，后直开领开大 3cm，画顺新领口线。

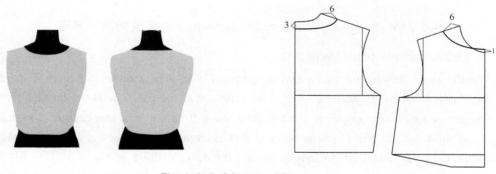

Fig. 4-2-2　A boat neckline
图 4-2-2 船形领口结构制图

3. A V-shaped neckline / V 形领

A V-shaped neckline with a welt is shown in Fig. 4-2-3. This shape is based on the basic neckline shape. There is no change on the basic front neckline width (if needed slightly wider, the neckline width can be increased by 0 to 1.5 cm). The depth of the front V-neckline is increased by 10cm; The back neckline depth and the front and back neckline widths are increased by 3 cm. Draw the new neckline and finish the welt according to the neckline style.

　　如图 4-2-3 所示，此领为装有贴边的 V 形领。在基础领口基础上，前横开领不变化（若需稍微开大，横开领开大量一般为 0 ～ 1.5cm）。前直开领加深 10cm；后直开领，前、后横开领各加大 3cm。画顺新领口弧线，根据领型作出贴边。

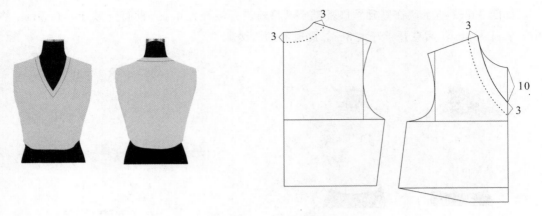

Fig. 4-2-3 A V-shaped necklines
图 4-2-3 V 形领结构制图

Unit 3 Pattern design for stand collars

第三节 立领纸样设计

一、Theory for designing stand collar patterns / 立领结构设计原理

1. Obtuse stand collars / 钝角立领

As the angle between the collar and the garment is an obtuse angle, this kind of stand collar is named obtuse stand collar, whose shape is like a frustum as the style line is shorter than the neckline. The pattern is initially developed in a rectangular shape from the front and back neck measurements. Changing the length of BB' (cocking up) will determine how the collar fits around the neck. If BB' is shorter, the collar edge will fit loosely around the neck (collar a in Fig. 4-3-1). If BB' is longer, the collar edge will fit tightly around the neck (collar b in Fig. 4-3-1). If BB' is longer to make the collar neckline is the same shape with the bodice neckline, the standing collar will disappear to be the extension from the body (collar c in Fig. 4-3-1). By adjusting the length of BB' (cocking up), various styles can be created. But the length of BB' is conditional, such as the length of the collar edge must be longer than the neck circumference. As shown in Fig. 4-3-1.

钝角立领指立领与衣身的角度呈钝角，立领的上领口线小于下领口线呈台体。该立领的样板基形是长度为前后领口弧线长的矩形。BB' 之间长度（起翘量）的变化，影响着领子贴合脖颈的状态：BB' 长度（起翘量）越小，领子上口线与脖颈之间的松量越多，越不贴合脖颈（图 4-3-1 中 a）；BB' 长度（起翘量）越大，领子上口线越紧密地贴近脖颈（图 4-3-1 中 b）。当领口曲线与领窝曲线完全吻合时，立领特征消失，变为衣身的延长（图 4-3-1 中 c）。通过改变 BB' 之间的长度（起翘量），可以创造出不同的钝角立领。然而领底线起翘的选择是有条件的，如领底线上翘应保证上口围度不能小于颈围。如图 4-3-1 所示。

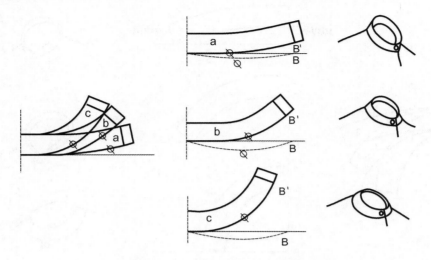

Fig. 4-3-1 Trends for obtuse stand collars

图 4-3-1 钝角立领变化趋势

The length of BB' (cocking up) is determined by the difference between the collar edge and the bodice neckline. BB' is generally 0-3 cm, usually 1.5-2.0 cm. The greater the length of BB', the more closely the collar is applied to the neck. When the amount of cocking up is large, it is necessary to open a large collar or choose a smaller collar band.

BB' 的长度（起翘量）是由领上口围和颈根围的差值确定。起翘量 BB' 一般为 0～3cm，常取 1.5～2cm。BB' 的长度（起翘量）越长，领子越贴向脖颈。起翘量较大时要适当开大领口或选择较小的领座。

2. Acute stand collars / 锐角立领

Opposite to the obtuse angle stand collar, the angle between the collar and the garment is an acute angle, so this kind of stand collar is named acute stand collar. Its structure principle is also opposite to obtuse angle stand collar. Changing the length of BB' will determine how far the collar band is away from the neck. When BB' is shorter, the collar edge will fit loosely around the neck (See collar a in Fig. 4-3-2). When the length of BB' is larger, the farther the collar edge is from the neck, the longer the collar edge is than the neckline, the more the collar inclines outward, and the easier it is to turn over the collar, which constitutes the structure of the collar band and fold in fact (See collar b in Fig. 4-3-2). When the bottom line of the collar matchs the neckline completely, the characteristics of the stand collar disappear, and it become a flat collar (See collar c in Fig. 4-3-2).

与钝角立领恰好相反，锐角立领指立领与衣身的角度呈锐角。其结构原理也与钝角立领相反。BB' 之间长度的变化，影响着领子远离脖颈的状态。BB' 长度较小时，领子上口线松散的围绕着脖颈（图 4-3-2 中 a）。BB' 长度较大时，领子上口线越远离脖颈，上领口线比领口线长得越多，领子越向外倾斜，越容易翻折，构成事实上的领座和翻领的结构（图 4-3-2 中 b）。当领下口弧线与颈根围弧线完全吻合时，立领特征消失，变为平贴领（图 4-3-2 中 c）。

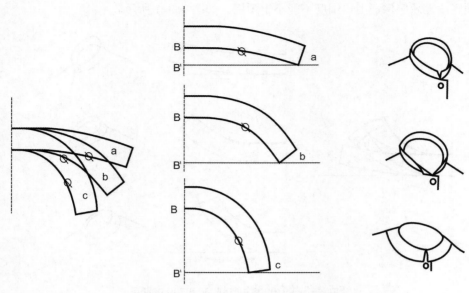

Fig. 4–3–2 Trends for acute stand collars
图 4-3-2 锐角立领变化趋势

二、Measuring the neckline / 测量领口

Measure the length of the neckline after adjusting it according to the style. The stand collar must be measured accurately with the tape measure pointed upright. Back-to-front placement of shoulders touching, measure along the neckline curve from the front centerline to the back centerline for 1/2 of the neckline curve length. As shown in Fig. 4-3-3.

先根据款式将领口线进行调整之后再测量领口线的长度。立领必须使用卷尺精确地垂直测量。重合前、后肩线，沿着领口弧线由前中心线测量至后中心线即为 1/2 领口弧线长。如图 4-3-3 所示。

三、Examples / 结构制图举例

1. Mandarin collar / 中式立领

(1) Measure the neckline/ 测量领口

Measuring the neckline as shown in Fig. 4-3-3.

测量领口如图 4-3-3 所示。

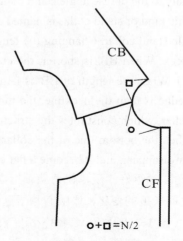

$○+□=N/2$

Fig. 4–3–3 Measure the neckline
图 4-3-3 测量领口

(2) Drawing Steps are shown in Fig. 4-3-4 / 制图步骤如图 4-3-4 所示：

① Draw rectangle: A-B = 1/2 of neckline measurement. AD = 3cm(collar band height). AD is the back center line of the collar. BC is parallel and equal to AD.

画长方形：AB 为 1/2 领围长，AD 为 3cm（立领高）。AD 为领后中心线。BC 与 AD 平行且相等。

② Find point F on BC to make FB=1.5cm. Divide AB into three equal sections by points E and H. Connect points E and F, and make FG perpendicular to EF, and its length is equal to BC.

在 BC 上找点 F，使 FB=1.5cm。将 AB 三等分，连接点 E、F，作 FG 垂直于 EF，且长度与 BC 相等。

③ Draw the outer line smoothly to complete the stand collar pattern design.

画顺外轮廓线，完成立领纸样设计。

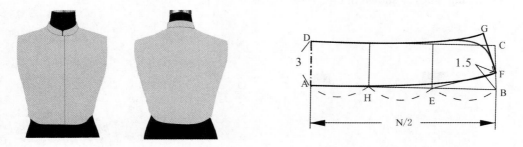

Fig. 4-3-4 The Mandarin collar
图 4-3-4 中式立领结构制图

2. Turndown collar / 翻立领

(1) Measure the neckline/ 测量领口

Measure the neckline as shown in Fig. 4-3-3.

测量领口如图 4-3-3 所示。

(2) Drawing Steps are shown in Fig. 4-3-5 / 制图步骤如图 4-3-5：

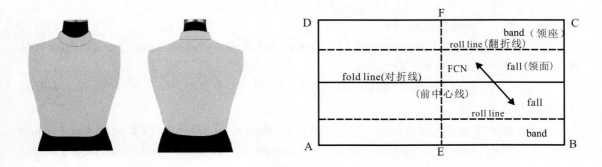

Fig. 4-3-5 The turndown collar
图 4-3-5 翻立领结构制图

① Draw a rectangle: AB = full neckline measurement; AD is twice the sum of the collar band height (3cm) and the fall depth (5cm).

画长方形：AB 为领围长；AD 为领座高（3cm）和领面高（5cm）之和的两倍。

② Mark the front center line, the roll line, the fold line.

标记前中心线、翻折线和对折线。

③ This type of collar is more suitable for cutting with knitted fabrics. If using a woven fabric, it should be bias-cut.

这种领型更适宜于用针织面料裁剪。如果是梭织面料则要斜裁。

3. Wing Collars / 翼领

(1) Measuring the neckline / 测量领口

Measuring the neckline as shown in Fig. 4-3-3.

测量领口如图 4-3-3 所示。

(2) Drawing Steps as shown in Fig. 4-3-6 / 制图步骤如图 4-3-6：

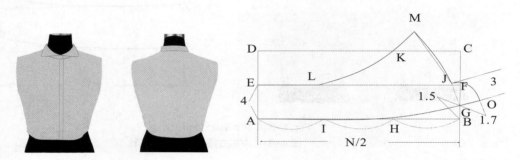

Fig. 4-3-6 The Wing collar
图 4-3-6 翼领结构制图

① Draw a rectangle ABCD: AB=N/2, AD=2×4cm (collar band height)=8cm; AD is the center line of the back collar; Point E is the midpoint of AD, and point F is the midpoint of BC, and connect points E and F.

作矩形 ABCD：AB=N/2，AD=2×4cm（领座高）=8cm；AD 为后领中心线；点 E 为 AD 的中点，点 F 为 BC 的中点，连接点 E、F。

② Find point G on BC to make GB=1.5cm. Divide AB into three sections. Connect points H,G. JG is perpendicular to HG and JG = 3cm.

在 BC 上找点 G 使 GB=1.5cm，将 AB 三等分。连接点 H、G。作 JG 垂直于 HG 且 JG=3cm。

③ Set point L on EF so that EL = 1/3 EF, and set point K on CD so that CK = 1/3 CD. Connect points E, L, and K to draw a curve, and then extend it to point M along the trend so that MJ = 1.5 CF.

在 EF 上定点 L，使 EL=1/3 EF；在 CD 上定点 K，使 CK=1/3 CD。弧线连接 E 点、L 点、K 点，并顺势延长至 M 点，使 MJ=1.5 CF。

④ Extend HG to point O, so that GO=1.7cm (Front overlap). Draw curves A-G-O, M-J, and J-O, as shown in Fig. 4-3-6.

延长 HG 至点 O，使 GO=1.7cm（前搭门量）。分别画顺弧线 A-G-O、M-J 及 J-O，如图 4-3-6 所示。

Unit 4　Pattern design for fold-over collars
第四节　翻折领纸样设计

A fold-over collar consists of a collar band part and a fold part. There are two types of fold-over collars. One is the ordinary fold-over collar, whose collar band is connected with the fold; another type is that the collar band part is separated from the fold part. The fold part on the suit and the lapel on the garment body join together and that is called the lapel collar. The lapel collar used on suits is a special type of fold-over collars. The collar band height (in this book, the default refers to the height of the back center) is generally 1–2 cm for the low collar band, 2.5–3 cm for the middle collar band, and 4–6 cm for the high collar band.

翻折领由领座和翻领两部分构成。翻折领主要分为两种类型，一种是普通翻领，其领座部分与翻领部分连在一起；另一种是领座部分与翻领部分分开的类型。西装上的翻领部分与衣身的驳头连在一起，称为翻驳领。西装的翻驳领属于翻折领的一种特殊类型。领座的高度（在本书中默认指后中的高度）一般为：低领座 1 ～ 2cm，中领座 2.5 ～ 3cm，高领座 4 ～ 6cm。

一、Theory of designing fold-over collar pattern / 翻折领结构设计原理

1. Experiment / 实验

The basic shape of the sample is a rectangle and its length is the sum of the front and back neckline curve lengths, and its width shall be the total width of the fold-over collar. If the fabric is cut out into a rectangle, being as a fold-over collar, sewed with the garment and rolled according to the roll line as the design, there will be a lot of wrinkles on the collar face. As shown in Fig. 4-4-1.

样板的基形是一个矩形，其长度为前、后领口弧线长度之和，宽为翻折领的总宽。若将面料裁剪成一块长方形，作为翻折领，将其与衣身缝合并按照设计的翻折线翻折，领面将会产生许多皱纹。如图 4-4-1 所示。

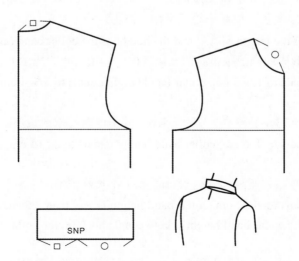

Fig. 4–4–1 The wrinkly fold-over collar
图 4-4-1 起皱的翻折领

If you take the side neck point as the center and make one auxiliary line for each side. Cut along the auxiliary lines separately to make those original wrinkles flatten out, and the cut will spread out naturally. Label these expanded gaps.

若以侧颈点为中心，每边各做一条辅助线。沿辅助线线分别剪开，使那些原本的皱褶变平展，剪开处会自然展开。标记这些展开的量。

Reshape the collar pattern, tracing around the gaps that marked above, so that the back center of the collar will rise and the shape of the collar outer edge will be more curved. As shown in Fig. 4-4-2.

在平面纸样上标出已加入展开量的尺寸，重新绘制图形，后领中心就会上升（产生了提高尺寸，领子外围线也成弧线状。如图 4-4-2 所示。

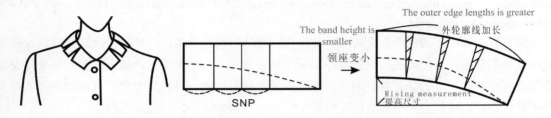

Fig. 4-4-2 Adjusting the fold collar
图 4-4-2 修正翻折领

2. Regularity/ 规律

As shown in Fig. 4-4-3, different neckline shapes form different collar outer edge lengths. As AA' is decreased, the bending downward degree of the lower neckline of the collar becomes weaker, resulting in a shorter collar outer edge, which causes the collar to stand up more. As AA' is increased, the bending downward degree of the lower neckline of the collar becomes stronger, resulting in a longer collar outer edge, which causes the collar to lie flatter on the shoulder.

如图 4-4-3 所示，不同的领口线形状形成不同的外轮廓线长度。当 AA' 减少时，领下口线向下弯曲的程度变弱，领外轮廓线变短，使领子更直立。当 AA' 增加时，领下口线向下弯曲的程度变强，领外轮廓线变长，使领子趋于平贴在肩上。

The change rule of the length of AA' and the height of the collar band (as shown in Fig. 4-4-3):

AA' 的长度与所形成的领子领座高的变化规律（如图 4-4-3 所示）：

(1) When AA' equals to 1.5-3 cm, collar band height equals to 3.5-4 cm, the collar will fit tightly around the neck.

当 AA' =1.5 ～ 3cm 时，领座高 =3.5 ～ 4cm，领子造型与颈部紧贴。

(2) When AA' equals to 4-6 cm, collar band height equals to 2.5-3 cm, and the collar is not close to the neck.

当 AA' =4 ～ 6cm 时，领座高 =2.5 ～ 3cm，领子不紧贴颈部。

(3) When AA' equals to 7-12 cm, collar band height is less than 2.5cm, and the collar is far away from the neck. When the collar band height is between 0 and 1cm, the collar can already be seen to flat ten on the shoulder.

当 AA' =7 ～ 12cm 时，领座高小于 2.5cm，领子远离颈部。领座高 =0 ～ 1cm 时，领子已经可以被看作是平贴在肩部。

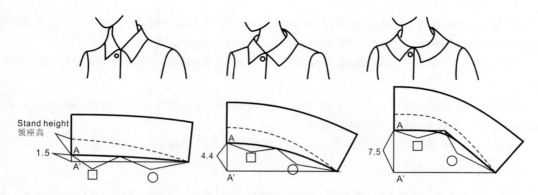

Fig. 4-4-3 Regularity for fold collars

图 4-4-3 翻折领变化规律

二、Examples / 结构制图举例

1. Overcoat Collars / 大衣领

The overcoat collar belongs to the ordinary fold collar whose collar band is connected with the fold。

大衣领属于领座与翻领部分连在一起的普通翻领。

(1) Measuring neckline / 测量领口

Trace around the bodice. Lower the neckline approx 1cm at the shoulders and 3cm at the front centerline. Mark points A, B, C, D on the new neckline. Measure the new neckline, mark AB=□ and CD=○. Mark in the button stand width(s) (3cm).

作出衣身结构图。前、后横开领开大 1cm，前中心降低 3cm。标注领口弧线上的 A、B、C、D 四点，测量前领口弧线 AB 的长度，标记为□，测量后领口弧线 CD 的长度，标记为○。标注搭门量（3cm）。

(2) Drawing Steps as shown in Fig. 4-4-4 / 制图步骤如图 4-4-4 所示：

① Make two perpendicular lines at point E. On the vertical line, GE=6cm, GI=3.5cm (collar band height), IJ=5.5cm (fall depth). At point G, GH is perpendicular to GJ and GH=○; HF=□, and point F is on the horizontal line. As shown in Fig. 4-4-4.

在 E 点作两条相互垂直的线。在竖直线依次向上量取 GE=6cm，GI=3.5cm（领座高），IJ=5.5cm（翻领高）。在 G 点作 GH 垂直于 GJ 且 GH=○，作 HF=□，F 在水平线上。如图 4-4-4 所示。

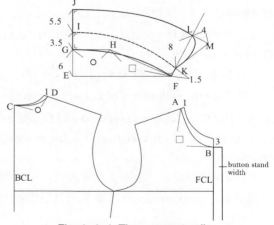

Fig. 4-4-4 The overcoat collar

图 4-4-4 大衣领结构制图

② Make the vertical line of HF, and measure FK=1.5cm and KL=8cm on it. Make LM perpendicular to KL and measure LM=4 cm on it. Connect points K, M.

作 HF 的垂线，在其上量取 FK=1.5cm，KL=8cm。作 LM 垂直于 KL，在其上量取 LM=4cm。连接点 K、M。

③ Through points G, L and K, draw the outer contour line of the collar. At the same time, draw the assembling neck line GF and roll line IK.

过点 G、L、K 画顺领子外轮廓线。同时，画顺领子的装领线 GF 和翻折线 IK。

2. Shirt collars / 衬衫领

The shirt collar belongs to the type that the collar band part is separated from the fold part. The collar band part is generally a single stand collar shape, and the fold part, according to the shape of the collar corner, is divided into pointed collar, small square collar, round collar and other forms.

衬衫领属于领座部分与翻领部分分开的类型。其领座部分一般为单立领造型，翻领部分根据领角的形态又分尖领、小方领、圆领等形态。

(1) Measure the neckline/ 测量领口

Measure the neckline as shown in Fig. 4-3-3.

测量领口如图 4-3-3 所示。

(2) Drawing Steps are shown in Fig. 4-4-5 / 制图步骤如图 4-4-5：

① Draw rectangle ABCD: AB=N/2, AD=3cm (collar band height), AD is the back center line.

作矩形 ABCD：AB=N/2，AD=3cm（领座高），AD 为后中心线。

② Extend AD to point E so that DE=2cm and continue to point F so that EF=4.5cm (the fold collar depth). Similarly, extend BC to point G and continue to point H so that CG=DE and GH=EF. Connect points E to G, F to H.

延长 AD 到 E 点，使 DE=2cm，继续延长至 F 点，使 EF= 翻领高 4.5cm。同理，延长 BC 到 G 点，继续延长至 H 点，使 CG=DE，GH=EF。连接 E 点和 G 点、F 点和 H 点。

③ Take the point K on BC so that KB=1.5cm. Bisect BC into 3 parts, mark equal points I and J, and connect points K and J. Make KM perpendicular to KJ and KM=2.5cm.

在 BC 上取点 K，使 KB=1.5cm。将 BC 平分成 3 份，标记等分点 I、J，连接 K 点和 J 点。作 KM 垂直于 KJ 且 KM=2.5cm。

④ Extend KJ to point L, so that KL=1.7cm (front overlap). Draw the curve A–K–L and the curve D–M–L separately.

延长 KJ 至 L 点，使 KL=1.7cm（搭门量）。分别画顺弧线 A-K-L 和弧线 D-M-L。

⑤ Take the midpoint N of EG and connect points N and M. Extend the straight line FH to point O, so that HO=2cm, connect points M and O, and extend it to point P, so that OP=0.6cm. Draw the outer contour line FP of the collar and the lower curve EM of the fold collar. The entire collar structure is completed.

取 EG 的中点 N，连接 N 点和 M 点。延长直线 FH 至 O 点，使 HO=2cm，连接 M 点和 O 点并延长至 P 点，使 OP=0.6cm。画顺领面外轮廓线 FP 和翻领下口弧线 EM。整个领子结构完成。

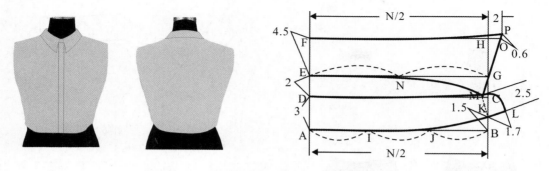

Fig. 4–4–5 Shirt Collar patterns
图 4–4–5 衬衫领结构制图

3. Classic lapel collars (Tailored) / 经典翻驳领

(1) Measuring the neckline / 测量领口

Make front and back bodice structures. Mark overlap (2 cm) in the front, according to the style drawing, mark point B as the break point (turn-over stop point). Mark the four points A, B, C and D on the neckline, measure the length of the back neckline curve CD, and mark as ☆.

作出前后衣身结构图。标注搭门量（2cm），按照款式图标注 B 点为驳点（翻折止点）。标注领口线上的 A、B、C、D 点，测量后领口弧线 CD 的长度，记为☆。

(2) Drawing Steps are shown in Fig. 4-4-6/ 制图步骤如图 4-4-6：

Extend the shoulder line to point E, so that AE=2cm, connect points B, E with a dotted line, BE is the roll line.

延长肩线到点 E，使 AE=2cm，用虚线连接 B、E，BE 即为翻折线。

Make the style line of the collar according to the effect drawing. The style line is symmetrical to the other side with the fold line as the axis of symmetry.

按照效果图作出领子的款式造型线。以翻折线为对称轴将造型线对称到另一侧。

Extend BE to point F, make AG parallel to BF, and AG= ☆ . GH is perpendicular to AG, and HG=3cm. Connect points A and H, and find point I on the line so that AI= ☆ . Make JI perpendicular to AI so that IP=2.5cm (collar band height) and PJ=4cm (fall height).

延长 BE 至 F 点，作 AG 平行于 BF，且 AG= ☆。作 GH 垂直于 AG，且 HG=3cm。连接 A 点和 H 点，在线上找到点 I，使 AI= ☆。作 JI 垂直于 AI，使 IP=2.5cm（领座高），PJ=4cm（翻领高）。

Draw the outer contour line JK of the collar. Extend the straight line LM to point N, so that NM=3cm. Curve IN as the assembling neck line. At the same time, draw the roll line PB. Mark O is the opposite point of assembling neck line and neck line. Sketch out the structure of the collar.

圆顺领面外轮廓线 JK。延长 LM 至点 N，使 NM=3cm。圆顺装领线 IN，同时画顺领子的翻折线 PB。标记点 O 为装领线和领口线的对位点。拓出领子的结构图。

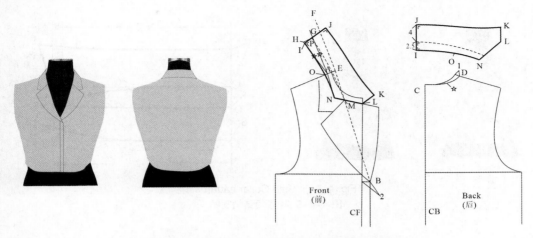

Fig. 4-4-6 Classic lapel collar (Tailored) patterns
图 4-4-6 经典翻驳领结构制图

4. Double-breasted lapel collars / 双排扣翻驳领

(1) Measure the neckline / 测量领口

Make front and back bodice structures. Mark in the front overlap (8cm), according to the style drawing, mark point B as the break point (turn-over stop point). Mark the four points A, B, C and D on the neckline, measure the length of the back neckline curve C and D, and mark as ☆.

作出前后衣身结构图。标注搭门量（8cm），按照款式图标注 B 点为驳点（翻折止点）。标注领口线上的 A、B、C、D 四点，测量后领口弧线 CD 的长度，记为☆。

(2) Drawing Steps as shown in Fig. 4-4-7/ 制图步骤如图 4-4-7:

Extend the shoulder line to point E, so that AE=2cm, connect the BE with a dotted line, BE is the roll line.

延长肩线到点 E，使 AE=2cm，用虚线连接 BE，BE 即为翻折线。

Make the style line of the collar according to the effect drawing. The style line is symmetrical to the other side with the fold line as the axis of symmetry.

按照效果图作出领子的款式造型线。以翻折线为对称轴将造型线对称到另一侧。

Extend BE to point F, make AG parallel to BF, and AG= ☆ . GH is perpendicular to AG, and HG=5cm. Connect points A and H and find point I on the line so that AI= ☆ . Make JI perpendicular to AI so that IP=3cm (collar band height) and PJ=5cm (fall height). .

延长 BE 至 F 点，作 AG 平行于 BF，且 AG=☆。作 GH 垂直于 AG，且 HG=5cm。连接 A 点和 H 点，在线上找到点 I，使 AI=☆。作 JI 垂直于 AI，使 IP=3cm（领座高），PJ=5cm（翻领高）。

Draw the outer contour line J-K of the collar. Extend the straight line LM to point N, so that NM=3cm. Curve I-N as the assembling neck line. At the same time, draw the roll line P-B. Mark O is the opposite point of assembling neck line and neck line. Sketch out the structure of the collar.

圆顺领面外轮廓线 J-K。延长 LM 至点 N，使 NM=3cm。画圆顺装领线 I-N，同时画顺领子的翻折线 P-B。标记点 O 为装领线和领口线的对位点。拓出领子的结构图。

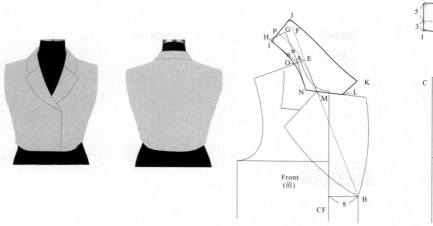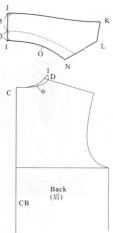

Fig. 4–4–7 Double–breasted lapel collar patterns
图 4–4–7 双排扣翻驳领结构制图

5. Shawl collars / 青果领

(1) Measure the neckline / 测量领口

Make front and back bodice structures. Mark in the front overlap (2cm), according to the style drawing, mark point B as the break point (turn-over stop point). Mark points A, B, C and D on the neckline, measure the length of the back neckline curve C and D, and mark as ☆.

作出前后衣身结构图。标注搭门量（2cm），按照款式图标注 B 点为驳点（翻折止点）。标注领口线上的 A、B、C、D 点，测量后领口弧线 CD 的长度，记为☆。

(2) Drawing Steps as shown in Fig. 4-4-8/ 制图步骤如图 4-4-8：

Extend the shoulder line to point E, so that AE=2cm, and connect points B, E with a dotted line. BE is the roll line.

延长肩线到点 E，使 AE=2cm，用虚线连接 BE，BE 即为翻折线。

Make the style line of the collar according to the effect drawing. The style line is symmetrical to the other side with the fold line as the axis of symmetry.

按照效果图作出领子的款式造型线。以翻折线为对称轴将造型线对称到另一侧。

Extend BE to point F, make AG parallel to BF, and AG= ☆ . GH is perpendicular to AG, and HG=4cm. Connect points A and H and find point I on the line so that AI= ☆ . Make JI perpendicular to AI so that IL=3cm (collar band height) and LJ=4.5cm (fall depth).

延长 BE 至 F 点，作 AG 平行于 BE，且 AG= ☆。作 GH 垂直于 AG，且 HG=4cm。连接 A 点和 H 点，在线上找到点 I，使 AI= ☆。作 JI 垂直于 AI，使 IL=3cm（领座高），LJ=4.5cm（翻领高）。

Draw the outer contour line JB of the collar. Curve IB is the assembling neck line. At the same time, draw the roll line LB. Mark K is the opposite point of the assembling neck line and neck line. Sketch out the structure of the collar.

圆顺领面外轮廓线 JB。画圆顺装领线 IB，同时画顺领子的翻折线 LB。标记点 K 为装领线和领口线的对位点。拓出领子的结构图。

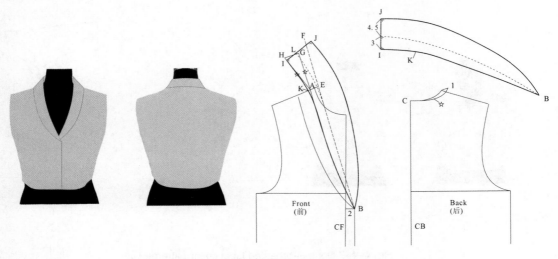

Fig. 4–4–8 Shawl collar patterns
图 4–4–8 青果领结构制图

Unit 5 Pattern design for developing collars
第五节 变化领纸样设计

一、Draping collars / 垂褶领

1. Measure the neckline / 测量领口

Lower the neckline approx 4 cm at the front centerline. Mark the new neckline points A, B. Measure the new neckline.

领口前中心降低约 4cm，标出新领口线上的 A、B 两点，测量新领口弧线的长度。

2. Drawing Steps as shown in Fig. 4–5–1/ 制图步骤如图 4–5–1：

(1) Mark three curves DE, FG, HI according to dividing lines to create a wider shape. Cut along these curves and pull out a certain pleat volume respectively. The pleated width in shoulder line is 3–5cm, and that in front centerline is 5–7cm.

在衣身上按照款式造型作出垂褶造型分割曲线 DE、FG、HI，并沿这些分割曲线剪开，拉展出一定的褶量。在肩线上的褶量为 3 ～ 5cm，在前中心线处的褶量为 5 ～ 7cm。

(2) The curve AB of the neckline has been changed to a new position. Point A goes to the position of point J. Draw a circle with point J as the center and the length of arc AB as the radius. Make the tangent line of the circle from C, and the tangent point is K. Connect J to K, JK=AB. The length of DJ is equal to AD plus pleats, and the length of CK is equal to CB plus drapes. In order not to affect the picture, delete the circle.

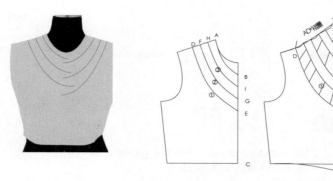

Fig. 4-5-1 Draping collar patterns

图 4-5-1 垂褶领结构制图

领口弧线 AB 变化后到了新的位置。A 点到了 J 点的位置。以 J 点为圆心，AB 弧线长为半径画圆。从 C 点做圆的切线，切点为 K 点。连接 J 点和 K 点，JK=AB。DJ 的长度等于 AD+ 褶量，CK 的长度等于 CB+ 垂褶量。为不影响看图，删掉圆。

(3) Extend KC to point C' so that CC'=1.5–2cm. Draw a new hem curve from C' point to the lower point of the side seam so that it is perpendicular to C'k. This is half the collar, and C'k is the new front centerline and symmetry line.

延长 KC 至 C' 点，使 CC'=1.5-2cm，从 C' 点向侧缝下点画顺新的下摆弧线使之与 C'K 垂直。这是一半的领子，C'K 是新的前中心线和对称线。

(4) Cowl neckline is suitable for bias-cut fabric.

垂褶领适宜于斜裁面料。

二、Seamless collars / 连身立领

1. Measure the neckline / 测量领口

Lower the neckline approx 2 cm at the shoulders, 0.5 cm at the back centerline, and 2.5 cm at the front centerline. Mark points A, B, C, D on the new neckline. Measure the new neckline.

开大领口：在侧颈点约 2cm，后中降低 0.5cm，前中心降低 2.5cm。标出新领口弧线上 A、B、C、D 点，测量新领口弧线的长度。

2. Drawing Steps as shown in Fig. 4-5-2 / 制图步骤如图 4-5-2：

(1) Extend the back centerline to point E so that DE=a=4cm (collar band height). The collar band height can be changed according to different styles.

延长后中心线至点 E，使 DE=a（领座高 4cm），领座高数值可以根据不同款式而变化。

(2) Draw a vertical line from point C to point F so that CF=3cm; make GF perpendicular to CF so that GF=1.2cm; connect points C and G, and extend it to point H, so that CH=a- 0.3cm.

由 C 点作竖直线至点 F 使 CF=3cm，作 GF 垂直于 CF 使 GF=1.2cm，连接 C 点与 G 点并延长至 H，使得 CH=a-0.3cm。

(3) Smooth the HC and shoulder line with a curve. Draw a curve from point H to point E and extend it to point I so that EI=0.3–0.5cm. Connect D and I. Mark the Stretch symbol (0.5–1cm) on the upper line HI of the collar.

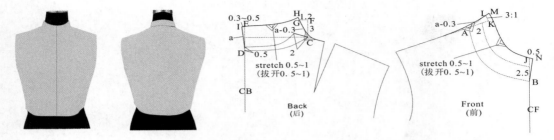

Fig. 4-5-2 Seamless collar cut in one with garment

图 4-5-2 连身立领结构制图

用弧线修顺 HC 和肩线。从 H 点向 E 点画弧线，并延长至点 I 使 EI=0.3 ～ 0.5cm。连接 D 点和 I 点。在领子上口线 HI 上标注拨开符号（拨开 0.5 ～ 1cm）。

(4) Extend the front centerline to point J and BJ=a.

延长前中心线至点 J，BJ=a。

(5) Extend the shoulder line to point K and AK=3cm. Make LK perpendicular to AK and LK=1cm. Connect points A, L and extend it to point M, and AM=CH=a-0.3cm.

延长肩线至点 K，AK=3cm。作 LK 垂直于 AK 且 LK=1cm，连接 AL 并延长至点 M，使 AM=CH=a -0.3cm。

(6) Trace around the shoulder line to point M, and point M to J with a smooth line. Extend curve MJ to point N and JN=0.5cm. Connect points B and N. Mark a stretch mark in the center of MN (stretch 0.5–1 cm).

用弧线画顺肩线和 MJ。延长 MJ 至点 N（JN=0.5cm）。连接 B、N 点并画顺领子外轮廓线。在领子上口线 MN 中心处标注拨开符号（拨开 0.5 ～ 1cm）。

三、Flat Collars / 平领

1. Drawing Steps as shown in Fig. 4-5-3 / 制图步骤如图 4-5-3：

(1) Place the shoulder of the back bodice to shoulder of the front bodice, with the side neck points touching, the shoulder endpoints overlap by 3 cm (the measurement can be changed with different styles).

将前后衣身侧颈点重合，肩端点重叠 3cm（可以依据款式而变化）。

(2) CF=AG=10cm. Draw the shape of the collar according to the style. Copy the outline of the collar. This is half of the collar, where CF is its back center line and line of symmetry.

量取 CF=AG=10cm。根据款式图画出领子的形状，描下领子轮廓。这是一半的领子，其中 CF 是其后中心线和对称线。

2. Knowledge Points / 知识点

A flat collar is actually a special fold collar with its stand height less than 1.5 m. When making the pattern, the overlapping value of the front and back shoulder lines and the height of the collar band influence each other, the more the shoulder lines overlap, the wider the stand depth.

平领实际上是一种领座高度小于 1.5cm 的特殊翻折领。在制作纸样时，前后肩线重叠的值和领座的高度相互影响，前后肩线重叠得越多，领座的高度越大。

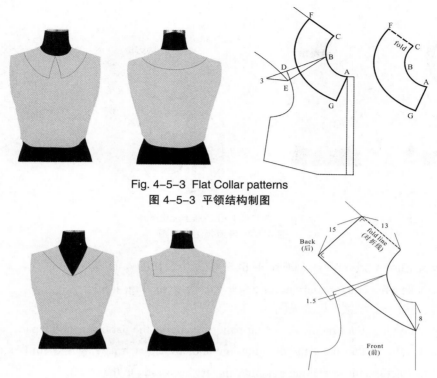

Fig. 4-5-3 Flat Collar patterns
图 4-5-3 平领结构制图

Fig. 4-5-4 Sailor Collar patterns
图 4-5-4 海军领结构制图

(1) If the shoulder lines overlap by 1cm, the collar band height is almost 0 cm.
当重叠量为 1cm 时，形成几乎没有领座的领子。

(2) If the shoulder lines overlap by 2.5cm, the collar band height is almost 0.6 cm.
当重叠量为 2.5cm 时，领座高为 0.6cm 左右。

(3) If the shoulder lines overlap by 3.8cm, the collar band height is almost 1 cm.
当重叠量为 3.8cm 时领座高为 1cm 左右。

(4) If the shoulder lines overlap by 5cm, the collar band height is almost 1.3 cm.
当重叠量为 5cm 时，领座高为 1.3cm 左右。

The outline of the collar can be drawn according to the collar style. Another "stand-less" collar called Sailor Collar is shown in Fig. 4-5-4.
领子外轮廓线的形状根据具体款式绘制。图 4-5-4 所示是另外一款低领座的领子，叫做海军领。

四、Frilled Collars / 波浪领

1. Measure the neckline / 测量领口

Lower the neckline by approx 6cm at the front centerline (The size of the neckline can be changed according to the style). Mark the new neckline points A, B. Measure the new neckline.
开大领口，前中心降低约6cm（领口改变的尺寸可以根据款式而变化），标记新领口线上的A、B 点，测量新领口弧线的长度。

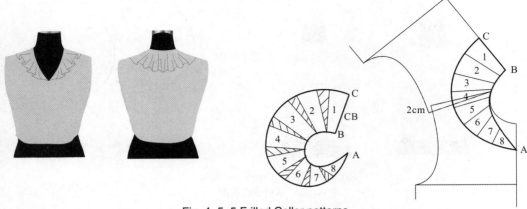

Fig. 4–5–5 Frilled Collar patterns
图 4-5-5 波浪领结构制图

2. Drawing Steps are shown in Fig. 4–5–5/ 制图步骤如图 4-5-5

(1) First make a flat collar pattern according to the style (see Fig. 4-5-3).

先根据款式做一个平领样板（见图 4-5-3）。

(2) Draw 7 dividing lines on the flat collar pattern to divide the flat collar into 8 parts, and cut the collar pattern. Cut the collar along the dividing line and stretch each part by 4–5 cm (As the shaded part shown in the picture, the stretching amount can vary according to the style).

在平领样板上画 7 条分割线，将平领分成 8 部分，并剪下领子样板。沿分割线剪开领子并展开，每个的拉展量为 4 ～ 5cm（如图阴影部分，拉展量可以根据款式变化）。

(3) Draw the collar outline smoothly, and trace the collar pattern. The finished part is the half collar pattern with BC as the back centerline.

画顺领子外轮廓线，拓下领子样板。完成的是半个领子的样板，BC 为其后中心线。

Unit 6 Making patterns of developing collars with CAD software

第六节 用 CAD 软件制作变化领纸样

There are many kinds of collar styles, and it is nothing more than a few basic collar models and the basic model based on the combination of darts, pleats and other changes out of collar models. Using CAD software to make these collar patterns is very simple.

领子款式有许多种，不外乎就是几种基本领型和在基本型基础上结合省道、褶裥等变化出来的领型。利用 CAD 软件制作这些领型样板都很简单。

一、Stand collars / 立领

The operation steps are as follows / 操作步骤如下：

Make a rectangle from the collar circumference and neck width – take out the rectangle into a pattern with the "Take out pattern" tool – make a few auxiliary lines – use the tool of "cut and move pattern", and make the top neckline curve shorter by a certain amount (input a negative number, which can be adjusted according to the needs of the style) – redraw the top and bottom collar lines – replace the original top and bottom collar lines with the tool of "Replacement net outline". As shown in Fig. 4-6-1.

由领围和领宽作长方形—用"取样片"工具将长方形取出成为样片—作几条辅助线—利用"样片剪开移动"工具，使上领口弧线缩短一定量（输入负数，可根据款式需要进行调整）—将领上口线和领下口线重新画顺—用"样片换净边"工具替换原来的领上口线和领下口线。如图 4-6-1 所示。

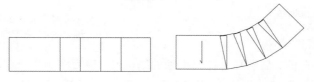

Fig. 4-6-1 Pattern making of stand collars
图 4-6-1 立领制作

二、Fold collars / 翻折领

In contrast to the standing collar, as shown in Fig. 4-6-2, use the " Cut and Move pattern" tool to lengthen the top line of the folding collar by a certain amount (input positive number, can be adjusted according to the style).

与立领相反，如图 4-6-2 所示，用"样片剪开移动"工具将翻折领的领上口线加长一定数量即可（输入正数，可根据款式需要调整）。

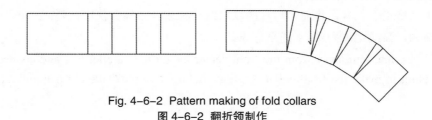

Fig. 4-6-2 Pattern making of fold collars
图 4-6-2 翻折领制作

三、Flat Collars / 平领

Flat collar pattern can be directly made on the front and back pieces after the shoulder seam is sewn, or a flat collar pattern can be made on the front and back pieces after the shoulder seams overlap slightly at the shoulder point, as shown in Fig. 4-6-3. After the shoulder seams of the front and back pieces overlap 1/4 of length of shoulder line at the shoulder point, make the structure drawing of the flat collar on the front and back pieces. Finally, take out the structure drawing into pattern. Its shape and size may vary according to the design, like a sailor collar as shown in Fig. 4-6-4.

平领样板可直接在肩缝拼合后的前后衣片上制作，也可使前、后片的肩缝在肩端点处少量重叠后再在前后衣片上作出平领的样板，如图 4-6-3 所示。将前、后片的肩缝在肩端点处重叠

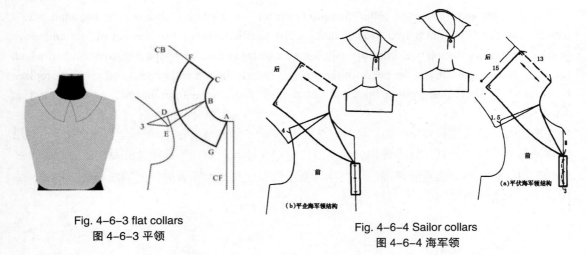

Fig. 4-6-3 flat collars
图 4-6-3 平领

Fig. 4-6-4 Sailor collars
图 4-6-4 海军领

1/4 肩线长的量后，再在前、后衣片上作出平领的结构图。最后将领片结构图取出成样片即可。平领的尺寸和形状也可以根据具体款式进行调整，如图 4-6-4 所示的海军领。

Copy the back and front bodice/ 复制前、后片：

① Take a point 3 cm below the shoulder point on the front armhole curve and connect it with the side neck point. Use the "Align" tool to select the entire back piece so that the shoulder line of the back piece is aligned with the connecting line on the front piece, as shown in Fig. 4-6-5.

在前袖窿弧线上肩点下方 3cm 处取一点，将它与侧颈点连接成线。用"对齐"工具将整个后片选中，使后片的肩线与前片上的连接线对齐，如图 4-6-5 所示。

② Using the "Split Correction" tool, merge the necklines of the front and back pieces into one line and smooth the curve. Then use the "Parallel Line" tool to make a parallel line with a distance of 10 cm from the new neckline curve.

用"拼合修正"工具，将前、后片的领口线合并成一条线并修顺此弧线。再用"平行线"工具作与新领口弧线距离为 10cm 的平行线。

③ Take a point 4cm away from the front centerline on the parallel line and connect it with the front neck point to make the front collar Angle modeling. Use the "take out pattern" tool to take the

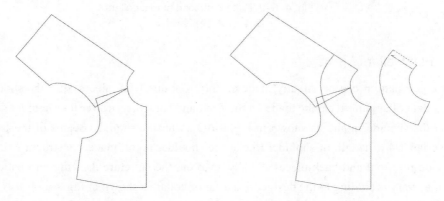

Fig. 4-6-5 Join the back and front bodice
图 4-6-5 拼接前、后片

Fig. 4-6-6 Take out the flat collar
图 4-6-6 取出平领样片

structure of the piece into a pattern. Select "Handle pattern – Symmetry – Symmetry Unwrap" to expand the collar, and then use "Symmetry – Symmetry Closure" to set the collar to be symmetrical.

在平行线上距前中心线 4cm 处取一点，与前颈点相连，作出前领角造型。使用"样片取出"工具将领子的结构图取成样片。选择"样片处理—对称—对称展开"来展开领片，然后使用"对称—对称闭合"来设置领为对称的。见图 4-6-6。

四、Cowl collars / 垂褶领

A cowl collar needs to increase the amount of pleats on the shoulders and the depth of the neckline to meet the drape effect, as shown in Fig. 4-6-7.

垂褶领需要在肩部和领深部位增加褶量，以满足褶皱效果，如图 4-6-7 所示。

The steps are as follows/ 操作步骤如下：

① Copy the front piece, adjust the front neckline curve to AB, draw three auxiliary lines as needed, and the lowest auxiliary line is marked CD. The bottom line is marked NE and mark the shoulder point as point P. Use the "Take out pattern" tool to take out the front piece into a pattern (including auxiliary lines), as shown in Fig. 4-6-8.

复制前衣片，调整前领口弧线为 AB，根据需要画几条辅助线，最低的辅助线标记为 CD，底边线标记为 NE，并标记肩端点为 P。用"样片取出"工具将前片取出成为样片（包括辅助线），如图 4-6-8。

② Click the "Cut and Move pattern" tool – select three auxiliary lines (in counterclockwise order) – right-click – input 3cm for "first end" and 6cm for "second end" in the dialog box – click "determine". The changes are shown in Fig. 4-6-7 (a).

点击"样片剪开移动"工具—选择 3 条辅助线（逆时针方向）—右键单击—在对话框中，"第一端"输入 3cm，"第二端"输入 6cm—点击"确定"。变化如图 4-6-7（a）所示。

③ With the "empty state" tool, place the mouse on the yarn center of the pattern, right-click, and click "pattern disassemble" to make the pattern become a structure drawing again. Take the new position of the side neck point A' as the center of the circle and the length of the arc AB as the radius to draw a circle. If the tangent line to the circle from the lowest point E of the front centerline is EF, connect A' to F, then A'F= arc length AB. Then delete the circle. As shown in Fig. 4-6-7(a).

用"空状态"工具将鼠标放在样片的纱向中心位右击，点击"样片解散"，使样片又成为结构图。以侧颈点所处新位置 A' 为圆心，AB 弧线长为半径画圆。从前中心最低点 E 做圆的切线 EF，连 A'F，则 A'F= 弧长 AB。再删掉圆，如图 4-6-7（a）所示。

④ Trim the shoulder line from point A' to point P. Extend FE to point M to make EM =1.5cm. Draw an curve from the lowest point N of the side seam to the point M so that FM is perpendicular to MN. As shown in Fig. 4-6-7(a).

从点 A' 到点 P 修顺肩线。延长 FE 到 M 使 EM=1.5cm，从侧缝下点 N 向 M 点画弧线，使 FM 垂直 MN。如图 4-6-7（a）所示。

⑤ Use the "Take out pattern" tool to take the changed outline as pattern. Curve A'P= shoulder length AP+ shoulder pleats , A'F=AB curve length, FM = front middle line BE+ drape folds. Mark the position of the shoulder pleats. As shown in Fig.4-6-7(b). Its shape and size may vary according to the design.

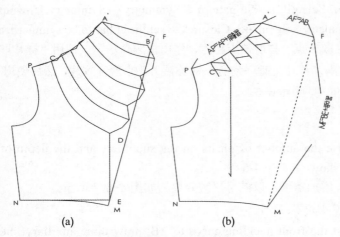

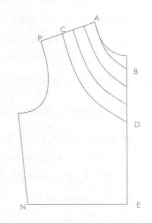

Fig. 4-6-7　Patterns of cowl collars
图 4-6-7　垂褶领结构图

Fig. 4-6-8　Draw auxiliary lines
图 4-6-8　作垂褶辅助线

　　用"样片取出"工具将变化后的轮廓取为样片。弧线 A' P= 肩线长 AP+ 肩褶量，A' F=AB 弧长，FM= 前中线 BE+ 垂褶量。标记肩线上褶的位置。如图 4-6-7（b）所示。垂褶领的形状和尺寸也可根据设计进行变化。

五、Others / 其他

There are also some by a few basic collars combined with darts, pleats and other forms of fashion collar model, and here is not an explanation. Shirt collar and lapel collar are introduced in Chapter 6, and their shapes and sizes can be changed according to the design.

　　还有一些在几种基本领型基础上结合省道、褶裥等形成的变化款领型，这里不一一讲解。衬衫领、驳领在第六章有介绍，其形状及尺寸可以根据需要变化。

Exercise/ 练习

1. Design 3 or 4 styles of collarless and make their patterns.

设计 3 或 4 款无领款式并制作它们的样板。

2. Make patterns of classic shirt collar.

制作经典衬衫领样板。

3. Make patterns of classic men's lapel Collar (Tailored) .

制作经典翻驳领样板。

4. Design 3 or 4 styles of developing collar and make their patterns.

设计 3 或 4 款变化领型并制作它们的样板。

Chapter 5 Sleeve pattern design
第五章 袖子纸样设计

The sleeve is the cover that connects the body piece and wraps the arms. The collar is also an important part of garment sections that affects the garments appearance and style. This chapter mainly tells the principles and methods of making sleeve pattern according to the armhole curve of the garment body, except for the armhole curves, the other structures of the bodice are omitted.

袖子是指连接衣身袖窿并包覆手臂的覆盖物，是服装构成中的一个重要组件，直接影响服装的外貌和风格。本章主要讲述根据衣身的袖窿弧线配制各种袖子纸样的原理和方法，衣身除了袖窿弧线外其他结构都省略。

Unit 1 Basic knowledge of sleeve pattern design
第一节 袖子纸样设计基础知识

一、Types of sleeves / 袖子的种类

The style of sleeves can vary diversely. There are many ways to classify sleeves. In terms of length, there are long sleeves, short sleeves, middle length sleeves, seven-point sleeves and nine-point sleeves; According to the number of divided pieces of the sleeve, there are the one-piece sleeve, two-piece sleeve, three-piece sleeve and kimono sleeve; According to the shape, there are flare sleeves, puff sleeves, lantern sleeves, petal sleeves, bat sleeves, etc.; According to the changed parts, there are the sleeve whose sleeve-opening changes and the sleeve whose sleeve-crown changes, etc.

袖子的变化很多。袖子的分类方法很多。从长短上来分，有长袖、短袖、中袖、七分袖和九分袖；从袖子的片数来分，有一片袖、两片袖、三片袖及连身袖；从造型上来分，有喇叭袖、泡泡袖、灯笼袖、花瓣袖、蝙蝠袖等；从变化的部位来分，有袖口变化袖、袖山变化袖等。

二、Common nouns on sleeves / 袖子上常用名词

Common nouns on sleeves are: armhole curve, sleeve length, sleeve width, sleeve crown height, sleeve crown curve, sleeve midline, sleeve elbow line, sleeve seam line, sleeve opening line, sleeve cuff and so on.

袖子上常用名词有：袖窿弧线、袖长、袖肥（或袖宽）、袖山高、袖山弧线、袖中线、袖肘线、袖缝线、袖口线、袖克夫等。

三、Theory of designing sleeve patterns / 袖子纸样设计原理

1. The relationship between the sleeve and body / 袖子与衣身的关系

The sleeve is made up of the sleeve crown and sleeve body. After the sleeve body is sewn, the sleeve hole is formed, which connected with the armhole of the garment body (see Fig. 5-1-1). The depth and shape of armhole determine the appropriate shape of sleeve hole. The depth and shape of armhole is determined by the designed body ease. When the garment body is fitted, the armhole is generally shallow and round; When the garment body is loose, the armhole is deep and narrow. Several common forms of armhole are shown in Fig. 5-1-2. There is also a special form of armhole which is a square armhole, and the square armhole in clothing is relatively rare.

袖子由袖山和袖身组成。袖身缝合后形成袖窿眼，与衣身的袖窿连接（见图 5-1-1）。袖窿的深浅和形态决定了与之相配的袖窿眼的形态。袖窿的深浅和形态则根据衣身款式的宽松程度来决定。衣身合体时，一般袖窿较浅而圆；衣身宽松时，袖窿则深而尖窄。袖窿的几种常见形态如图 5-1-2 所示。还有一种特殊的袖窿形态是方袖窿，而方袖窿在服装上比较少见。

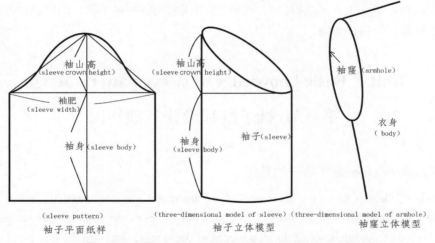

Fig. 5-1-1 The relationship between the sleeve and bodice body

图 5-1-1 袖子与衣身的关系

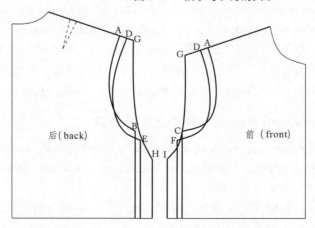

Fig. 5-1-2 Common forms of armholes

图 5-1-2 袖窿的几种常见形态

2. The relationship between the sleeve crown height and sleeve width / 袖山高与袖肥的关系

When the garment body fits, the armhole is usually shallow and round, which should match the fitted sleeve shape; when the body is loose, the armhole is deep and narrow, which should match the loose sleeve shape. What determines whether the sleeve is loose or not is the size of sleeve width.

衣身合体时，一般袖窿较浅而圆，与之相配的应该是合体的袖子造型；衣身宽松时，袖窿则深而尖窄，与之相配的就应该是宽松的袖子造型。而决定袖子宽松与否的则是袖肥的大小。

The height of the sleeve crown and the sleeve width are two important limits that affect the sleeve, these decide the shape and mutually restrict each other. For matching the same armholes, the higher the sleeve crown, the smaller the sleeve width and the more fitted the sleeve will be. In contrast, the lower the sleeve crown in height, the larger the sleeve width and the more ease there will be in the sleeve, as shown in Fig. 5-1-3.

袖山高和袖肥是影响袖子的两个重要参数，它们决定袖子的造型，彼此又相互制约着。对于匹配同一袖窿的袖子来说，袖山越高则袖肥越小，袖子越合体；相反，袖山越低则袖肥越大，袖子越宽松，见图 5-1-3。

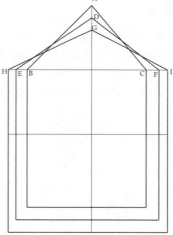

Fig. 5-1-3 The relationship between the sleeve crown height and sleeve width
图 5-1-3 袖山高与袖肥的关系

Generally speaking, the sleeve crown height taking between AH/4 minus 5 cm and AH/4 will form a low sleeve crown, and it is suitable for loose sleeves (such as sportswears); that taking between AH/4 and AH/4 plus 2.5 cm will form a middle sleeve crown, and it is suitable for common sleeves (such as leisure coats, summer shirts); that taking between AH/4 plus 2.5 cm and AH/4 plus 5 cm will form a high sleeve crown, and it is suitable for tight fitted sleeves (such as the male or the female suits).

一般来说，袖山高取值在 AH/4-5cm 至 AH/4 之间为低袖山，适用于宽松袖（如运动服）；取值在 AH/4 至 AH/4+2.5cm 之间为中袖山，适用于普通袖（如休闲外套、夏季衬衫）；取值在 AH/4+2.5cm 至 AH/4+5cm 之间为高袖山，适用于贴体袖（如男、女西服）。

But when the armhole cannot be connected to the sleeve; this kind of design becomes a sleeveless design.

然而衣身袖窿也可以不连接袖子。这种设计就变成了无袖设计。

Unit 2　Sleeveless design
第二节　无袖设计

A sleeveless design is essentially just an armhole design. This design is widely used in Summer clothing. If it is worn alone in summer, the armhole bottom of sleeveless clothing needs to be raised by 1.5–2 cm on basis of that of the prototype block in order to avoid "breast-baring". Meanwhile, the armhole and shoulder should be close to the arm root, so the shoulder dart and the breast dart should not be transferred into the armhole. If it is worn outside a shirt or sweater, the armhole bottom of sleeveless clothing needs to drop according to the design.

无袖设计也就是没有袖子的袖窿设计。无袖设计被广泛用于夏季服装中。若作为夏季单穿，其袖窿要在原型衣身的袖窿底基础上抬高 1.5 ～ 2cm，以免袖窿底"走光"。同时，袖窿和肩部也需贴近臂根，故后衣身肩省及前衣身胸省的量不能被转移到袖窿。若套在衬衫或毛衣外面穿，其袖窿底则要根据款式下落挖深。

一、Strapless sleeves / 露肩袖

When it's sleeveless, the armhole style can vary by moving the shoulder point. The shoulder point can be moved towards the neck and get the style that can be called the "strapless" sleeve as a result, as shown in Fig. 5-2-1, Fig. 5-2-2.

无袖时袖窿样式通过肩点移动可变化多样。肩点向颈部内移，就变成露肩袖，见图 5-2-1、图 5-2-2。

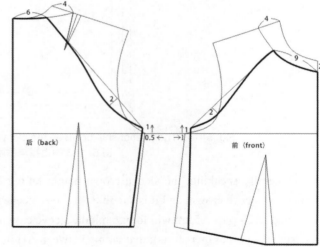

Fig. 5-2-1 The strapless sleeve
图 5-2-1 露肩袖款式图

Fig. 5-2-2 Patterns of strapless sleeves
图 5-2-2 露肩袖结构图

二、Dropped shoulder sleeves / 落肩袖

When the shoulder point moves outside, it can be called the "dropped shoulder" sleeve, as shown in Fig. 5-2-3, Fig. 5-2-4. Sleeveless armhole shapes can also be irregular depending on the style.

肩点外移则变成落肩，成为落肩袖，见图 5-2-3、图 5-2-4。无袖的袖窿形状也可以依据

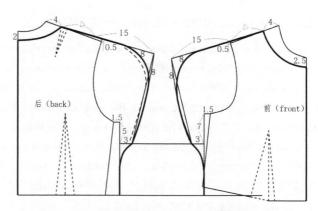

Fig. 5–2–3 The dropped shoulder sleeve
图 5-2-3 落肩袖款式图

Fig. 5–2–4 Patterns of dropped shoulder sleeves
图 5-2-4 落肩袖结构图

款式特点设计成不规则的。

Unit 3 Pattern design for sleeves with developing sleeve openings

第三节 变化袖口的袖子纸样设计

一、Straight sleeves / 直身袖

The straight sleeve opening is slightly smaller than the prototype sleeve opening, but after the sleeve is done its shape is straight, as shown in Fig. 5-3-1. The pattern processing is as follows: draw a sleeve prototype, narrow each side of the sleeve opening by 2 cm, then make the sleeve inside-seam concave by 0.5 cm at the elbow line, as shown in Fig. 5-3-2.

直身袖的袖口比原型袖口稍小，但袖子装好后形成了直筒形状，见图 5-3-1。纸样处理如下：描画原型袖，将袖口从两侧各缩小 2cm，将袖内缝线在肘弯处凹 0.5cm 并画顺，见图 5-3-2。

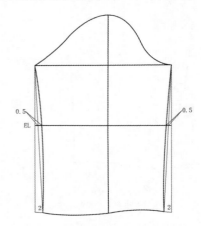

Fig. 5–3–1 The straight sleeve
图 5-3-1 直身袖款式图

Fig.s 5–3–2 The straight sleeve pattern
图 5-3-2 直身袖结构图

二、One-piece fitted sleeves / 一片合体袖

When there is a natural "droop" to the human arm's posture, the arm bends slightly forward, as shown in Fig. 5-3-3. The opening of one-piece fitted sleeve is comparatively smaller than the opening of prototype sleeve. When being completed, the sleeve will fit the curve of the arm, as shown in Fig. 5-3-4. Trace a sleeve prototype, shrink the sleeve opening from both sides and then angle the sleeve centerline forward 2–2.5 cm. The front sleeve opening size is the half sleeve opening size minus 1 cm. The back sleeve opening size is the half sleeve opening size plus 1 cm. The dart intake of sleeve elbow is the difference between the back and the front sleeve seams. As shown in Fig. 5-3-5(a).

人体手臂自然下垂时呈微向前弯曲的姿势，如图 5-3-3。一片合体袖的袖口相对原型袖的袖口更小。袖子成形后呈贴合手臂的弯曲状态，如图 5-3-4。将原型袖描画下来，从两边收小袖口，袖中线向前倾斜 2 ～ 2.5cm。前袖口大 = 袖口大 /2-1cm，后袖口大 = 袖口大 /2+1cm，袖肘省宽量是前后袖缝线的差量。如图 5-3-5(a) 所示。

Also, the dart of sleeve elbow can be transferred to the back sleeve opening, and then form the sleeve opening dart, as shown in Fig. 5-3-5(b).

此外，合体袖也可以把袖肘省转移至后袖口，形成袖口省。见图 5-3-5(b)。

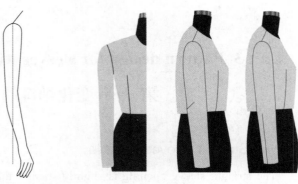

Fig. 5-3-3 Arm-placed forward
图 5-3-3 手臂前摆

Fig. 5-3-4 One-piece fitted sleeves
图 5-3-4 一片合体袖款式图

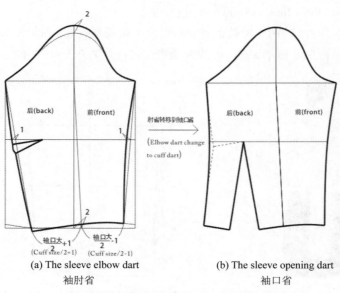

(a) The sleeve elbow dart
袖肘省

(b) The sleeve opening dart
袖口省

Fig. 5-3-5 Patterns of one-piece fitted sleeves
图 5-3-5 一片合体袖结构图

三、Flare sleeves / 喇叭袖

The name of flare sleeve is due to the "trumpet" shape, as shown in Fig.5-3-6. The patterns are designed as follows: trace the sleeve prototype and shorten its length to the elbow line. Draw some auxiliary lines on the sleeve pattern. Cut along the auxiliary lines and open these individual pieces respectively by 3 cm. The open amount should be distributed equally so that the flare shape is even. At the time the opening is enlarged, 2 cm ease was removed in the sleeve crown curve, as shown in Fig. 5-3-7.

喇叭袖因袖口呈喇叭状而得名，见图 5-3-6。纸样设计如下：将原型袖的袖长截至肘线位置，在袖子纸样上画几条辅助线，沿辅助线剪开并各展开 3cm，使袖口张开。展开的量要均匀分配，以使得喇叭形状均匀稳定。袖口放大的同时，袖山部位收掉吃势量 2cm，见图 5-3-7。

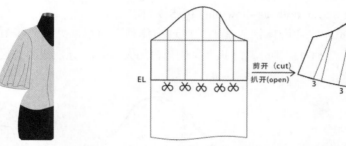

Fig. 5-3-6 The flare sleeve
图 5-3-6 喇叭袖款式图

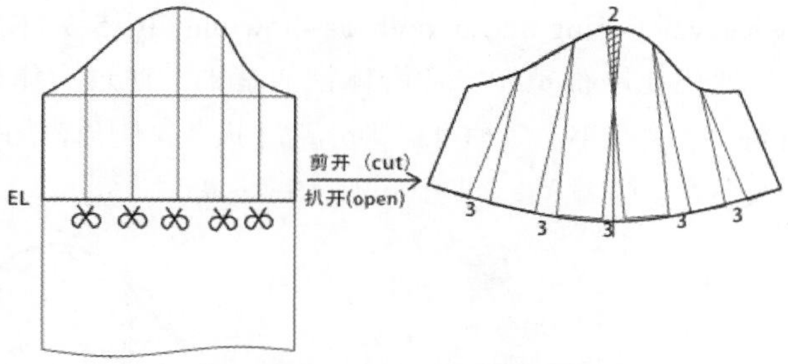

Fig. 5-3-7 Patterns of flare sleeves
图 5-3-7 喇叭袖结构图

四、Lantern sleeves / 灯笼袖

The sleeve is shaped like a Chinese lantern. It is rounder in shape in the middle and then smaller at the sleeve crown and the sleeve opening, as shown in Fig. 5-3-8. The pattern is designed as follows: First copy the sleeve prototype, shorten the sleeve length of the prototype sleeve as needed, and increase the sleeve crown height, and divide the sleeve piece into two parts at the largest part of the sleeve, and cut open the bottom of the upper part to make a trumpet shape. When the top of the lower part is cut open, its open size should be equal to the trumpet shape open size, then fold the lower part slightly and shrink the sleeve opening so that it can be distinguished from the largest part, as shown in

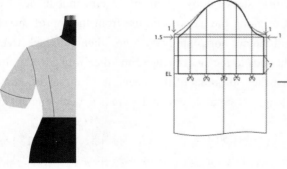

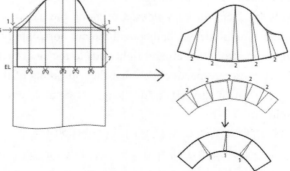

Fig. 5-3-8 Lantern sleeves
图 5-3-8 灯笼袖款式图

Fig. 5-3-9 Patterns of Lantern sleeves
图 5-3-9 灯笼袖结构图

Fig. 5-3-9.

灯笼袖是指上下收小而中间膨大的灯笼状袖型，见图5-3-8。纸样设计如下：先复制袖原型，根据需要将原型袖的袖长改短，并将袖山加高，在灯笼形最膨大的位置将袖分割成上下两部分。切开上部分的下端以做成喇叭形，把下部分的上端切开后使其张开的量与上部分喇叭形张开的量相同，然后把下端稍微折叠，以使袖口收紧，呈对比效果，见图5-3-9。

四、changed dress sleeves / 变化礼服袖

A changed dress sleeve is obtained by varying the sleeve opening on the basis of a one-piece fitted sleeve, as shown in Fig. 5-3-10. Trace the fitted sleeve, lengthen the sleeve crown height and reduce the sleeve width, split the back of the sleeve in half so that the sleeve opening is increased by 25 cm. On both sides of the cut line, design a slit strip placket, 8 cm long and 2.5 cm wide, then make the sleeve opening line smooth, as shown in Fig. 5-3-11.

变化礼服袖是在一片合体袖的基础上通过变化袖口而得，见图5-3-10。将合体袖的袖山加高、袖肥减小，后袖身对半分割并扒开，使袖口加大25cm；分割线两边设计长8cm、宽2.5cm的衩条，再圆顺袖口即可，见图5-3-11。

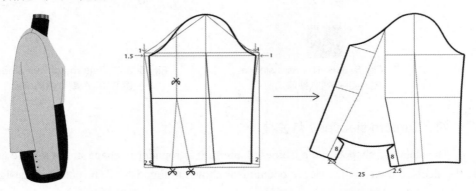

Fig. 5-3-10 Changed dress sleeves
图 5-3-10 变化礼服袖款式图

Fig. 5-3-11 Patterns of changed dress sleeves
图 5-3-11 变化礼服袖结构图

五、The sleeve opening tied / 袖口打结袖

The pattern design of the sleeve opening tied (as shown in Fig. 5-3-12) is created as follows: start with a basic sleeve prototype with a sleeve body length of 13 cm; indent the sleeve opening respectively by 2 cm on both sides. Cut along the sleeve midline from the top of the sleeve crown to the bottom line of sleeve, and spread the cut horizontally by 5 cm; then cut respectively along the sleeve width line from the left and right ends to the sleeve edges, and then rotate the splitted two parts of the sleeve crown respectively so that the top points of the sleeve is 13 cm apart, and draw a new smooth sleeve crown curve. The "tie" length is set to 27 cm, and there is a 5 cm long slit above the tie, as shown in Fig. 5-3-13.

袖口打结袖（见图5-3-12）的纸样设计如下：以原型袖为基础，袖身取13cm；袖口两边各缩进2cm。沿袖中线从袖山顶部往下剪开至袖底线，并水平拉开5cm；然后沿袖宽线分别从左右两端剪至袖边沿，再分别旋转被分割开的袖山两部分，使袖山顶点相距13cm，画顺新的袖山弧线。打结的长度设为27cm，其上端有一个5cm的开衩，见图5-3-13。

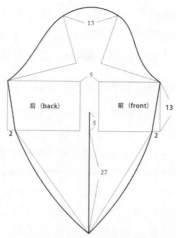

Fig. 5-3-12　The sleeve opening tied
图 5-3-12　袖口打结袖款式图

Fig. 5-3-13　Patterns of the sleeve opening tied
图 5-3-13　袖口打结袖结构图

Unit 4　Pattern design for sleeves with developing sleeve crowns

第四节　袖山变化的袖子纸样设计

一、Puff sleeves / 泡泡袖

The puff sleeve is the sleeve with puffy gathers, as shown in Fig. 5-4-1. When designing the patterns, choose the fitted sleeve block, and draw a horizontal cut line at the middle part of the sleeve crown. Cut vertically along the centerline to the horizontal line at the middle of the sleeve crown, and cut to both edges of the sleeve crown curve. Equally open both sides to create the desired style; make the sleeve crown curve coherently. Obtain the form of the sleeves "puff" by shrinking the increased amount of sleeve crown curve, as shown in Fig. 5-4-2。

泡泡袖是指袖山部位蓬松并有褶皱的袖型，见图 5-4-1。纸样设计时取合体袖的基本型，

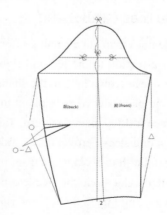

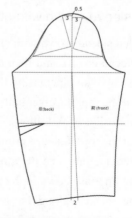

Fig. 5-4-1 Puff sleeves
图 5-4-1 泡泡袖款式图

Fig. 5-4-2 Patterns of puff sleeves
图 5-4-2 泡泡袖结构图

在袖山部位画一条水平剪切线，然后从上至下沿着袖中线剪开至水平线，并分别沿着这条线的左右两侧剪至袖山弧线。根据款式效果，分别向两边扒开一定的量，然后再连顺袖山弧线。缝制时将增加的袖山弧线量抽细褶，即得泡泡袖，见图 5-4-2。

二、The sleeve crown curve with darts / 袖山收省袖

The sleeve crown curve with darts refers to the sleeve type with darts along the sleeve crown curve, as shown in Fig. 5-4-3. The pattern design is as follows: Trace a sleeve block, reduce the sleeve width, and make the sleeve more fitted. Shorten the sleeve length by 14 cm, draw a horizontal line (i.e. dividing line) at the part of the sleeve crown and then draw three separate slanting lines that can cross the sleeve crown curve. Cut along these lines and open them, then draw the new sleeve crown curve and the three sleeve crown darts, as shown in Fig.5-4-4.

袖山收省袖是指沿袖山弧线收省的袖型，见图 5-4-3。纸样设计如下：描画合体一片袖纸样，将袖肥减小，使袖子更合体，袖长缩短 14cm。在袖山处画一水平分割线，再分别作三条斜线交于袖山，剪开水平分割线和斜线并扒开，将袖山弧线重新描好，做出三个袖山省，见图 5-4-4。

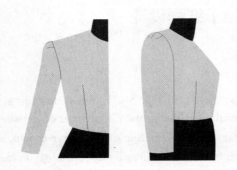

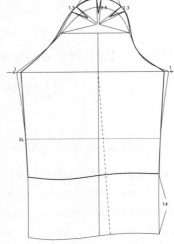

Fig. 5-4-3 The sleeve crown curve with darts
图 5-4-3 袖山收省袖款式图

Fig. 5-4-4 Patterns of the sleeve crown curve with darts
图 5-4-4 袖山收省袖结构图

三、The sleeve crown with style lines / 袖山分割袖

The so-called sleeve crown with style lines is to cut a small section along the sleeve midline from the top of the sleeve crown, and then cut a section to the left and right separately. By increasing the length of the cut curve and increasing the sleeve crown height, the shaped sleeve has a short horizontal standing effect at the top. It's essentially one-piece sleeve and not cut in two pieces, as shown in Fig.5-4-5. The pattern design is as follows: Trace a fitted sleeve pattern. Take 10 cm separately to the left and right sides at the top of sleeve crown along sleeve crown curve. Take the 20 cm long sleeve crown curve as the reference line, and draw a parallel line 3 cm apart from it to form a "style line". Draw a few more cutting lines as shown in the pattern. Cut along the sleeve centerline from the top point to the sleeve width line, then cut along the sleeve width line to the left and right endpoint (do not break), and expand it upward. Cut along the left and right style line at the top of the sleeve. From the above style line upward, cut along several lines on the left and the right respectively, and expand each cutline, so that the lengths of the style curves unfolded on the left and right are both 10 cm (marked as "o", "ø" respectively). Connect the left and right style curves below (marked as "□") and smooth the curve. If necessary, it can be slightly raised to make the curve □=o+ø +1cm (1 cm is the ease of this curve). As shown in Fig. 5-4-6.

所谓的袖山分割袖是通过在袖山顶部从袖中线处剪开一小段，再分别向左、向右剪开一段，通过增长剪开的弧线长和加高袖山，使得成型的袖子在顶部有一小段挺立的效果。其实质还是一片袖，并没被分割成两片，见图 5-4-5。纸样设计如下：描画合体一片袖纸样，在袖山弧线上从袖山顶点向左、右两侧各取 10cm。以此 20cm 的袖山弧线为基准线，画一条与它相距 3cm 的平行线，形成"分割线"。再按图中所示画几条剪开线。先沿着袖中线从顶点剪开至袖宽线，然后沿着袖宽线分别向左右两端剪开至端点（不要断开），再向上展开。最后剪切袖山顶部的左、右分割线。在上面的分割线上，左、右各剪开几条剪切线，展开各剪切线，使得左右展开后的分割弧线长（分别标为 o、ø）都为 10cm。连接下面的左右分割线（标为 □），修顺弧线。根据需要可略抬高，使弧线 □＝o +ø+1cm（1cm 即为这段弧线的缝缩量）。见图 5-4-6。

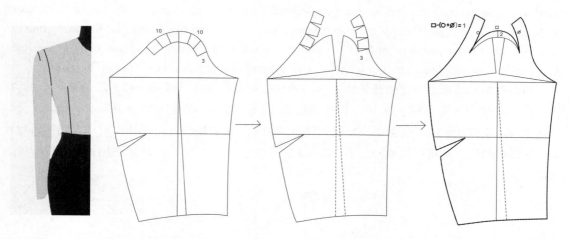

Fig. 5-4-5 The sleeve crown with style lines
图 5-4-5 袖山分割袖款式图

Fig. 5-4-6 Patterns of the sleeve crown with style lines
图 5-4-6 袖山分割袖结构图

四、Petal sleeves / 花瓣袖

Dividing a basic one-piece fitted sleeve into two "petal-shaped" pieces, as shown in Fig.5-4-7, creates the petal shape. The pattern design is as follows: cut the sleeve to the appropriate length, indent the sleeve opening by an appropriate amount on both sides to make the opening smaller. Draw the outline of the two petals according to the style. Cut out the pattern of the two petals, and then the two-piece petal sleeve is complete (as shown in Fig.5-4-8).

花瓣袖是在一片合体袖的基础上将袖子分割成花瓣状的袖型，见图 5-4-7。纸样设计如下：将一片合体袖截至适当的长度，分别在袖口两边缩进适当的量，使得袖口变小。根据款式图画出两片花瓣的轮廓线。将两片花瓣的纸样拓下来，即可得到两片袖型的花瓣袖（见图 5-4-8）。

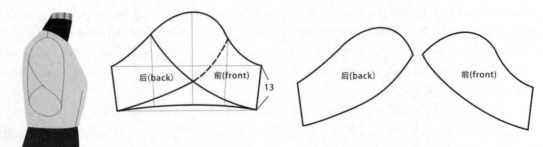

Fig. 5–4–7 Petal sleeves
图 5–4–7 花瓣袖款式图

Fig. 5–4–8 Patterns of Petal sleeves
图 5–4–8 花瓣袖结构图

五、The sleeve crown with a fold / 袖山折叠袖

The sleeve crown with pleat means that the sleeve crown is extended into a square and the sleeve body is folded forward to form a triangle sleeve shape, as shown in Fig. 5-4-9. The pattern design is as follows: firstly, the sleeve pattern length is shorten to 23.5 cm, the opening of sleeve and the sleeve width are reduced by 5 cm and by 3 cm respectively, and also the sleeve crown height is increased by 1cm, creating a tighter sleeve fit. Then separate the sleeve from the sleeve midline, draw line OA through point O, determine point B of the triangle on the sleeve crown curve and connect the lines. And finally draw the sleeve contour with AP as an symmetrical axis, as shown in Fig. 5-4-10.

　　袖山折叠袖是指袖山部位延伸成方形，袖身往前折叠形成三角形的袖子，见图 5-4-9。纸样设计方法如下：先将袖子原型长度缩短至 23.5cm，袖口和袖肥分别缩小 5cm 和 3cm，袖山抬高 1cm，使得袖子成为高袖山的窄袖。然后将袖子从袖中线分开，过袖山高点 O 延伸出线段 OA，在袖山弧线上确定三角形的一个点 B，连接各线。最后以 AP 为对称轴画出袖轮廓线即可，见图 5-4-10。

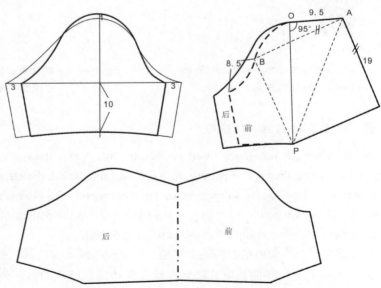

Fig. 5-4-9 The style of the sleeve crown with a fold
图 5-4-9 袖山折叠袖款式图

Fig. 5-4-10 Patterns of the sleeve crown with a fold
图 5-4-10 袖山折叠袖结构图

Unit 5 Two-piece sleeve pattern design

第五节 两片袖纸样设计

The tightness of one-piece fitted sleeves is limited, and it is difficult to get rid of the flat feeling. The two-piece sleeve is not only fitted but also three-dimensional, and it can evolve from the basic one-piece sleeve. The design pattern is as follows: trace the sleeve prototype, raise the sleeve crown 2 cm, fold the front and back sleeve in half, tilt the sleeve forward, and draw a smaller opening, cut along the sleeve side-seams and it becomes the two-piece sleeve. In order to make the sleeve more attractive, conceal the sleeve side-seams; divide the sleeve into the top and under sleeve pattern pieces. The top sleeve pattern piece is widened by 3 cm on the front and 1.5 cm on the back for covering the arm's front visible seams; The under sleeve is relatively hidden. As shown in Fig. 5-5-1, Fig. 5-5-2, Fig. 5-5-3 (Structure chart of two-piece sleeves with a slit).

The two-piece sleeve (front)
两片袖正面

The sleeve with a slit (back)
开衩两片袖背面

The sleeve without a slit (back)
不开衩两片袖背面

Fig. 5-5-1 Two-piece sleeves
图 5-5-1 两片袖款式图

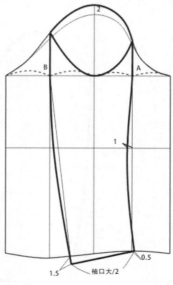
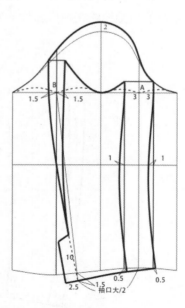

Fig. 5-5-2 One-piece sleeve becomes two-piece sleeve
图 5-5-2 一片袖变两片袖

一片袖做成的合体袖，合体程度是有限的，在外观上难以摆脱平面的感觉。而两片袖不仅合体还具有立体效果，它可在一片袖的基础上演变而来，具体做法是：描画袖原型，袖山加高 2cm，将前后袖对折，顺应手臂向前倾斜，得到袖口收小、袖身微向前倾斜的袖筒，剪开袖缝线即变成了两片袖。为了使袖子更为美观而将袖缝线隐蔽，袖子则分成大、小袖。大袖前后各借小袖 3cm 和 1.5cm，以覆盖手臂上可见的部分；小袖位于手臂内侧较为隐蔽的地方。见图 5-5-1、图 5-5-2、图 5-5-3（带袖衩的两片袖结构图）。

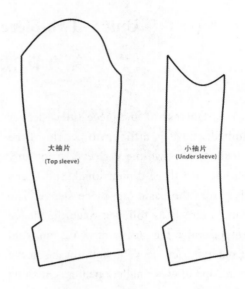

Fig. 5-5-3 Patterns of the two-piece sleeve with plackets

图 5-5-3 带袖衩的两片袖结构图

The two-piece sleeve has two forms: one is with a sleeve placket, the other is without a sleeve placket. The two-piece sleeve with sleeve placket has a sleeve placket at the sleeve opening, and the back sleeve seam of the top and under sleeves basically coincide together at the sleeve placket. For the two-piece sleeve without sleeve placket, the back sleeve seam of the top and under sleeves are separated at the opening of sleeve, as shown in Fig. 5-5-4.

两片袖有两种形式：一种是有袖衩的，另一种是没有袖衩的。有袖衩的两片袖在袖口部位有一个袖衩，其大、小袖片的后袖缝线在袖衩部位基本重合在一起。不带袖衩的两片袖，其大、小袖片的后袖缝线在袖口部位是分离的，如图 5-5-4 所示。

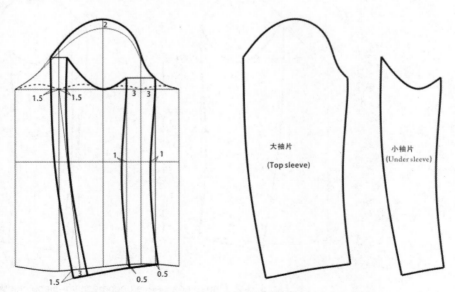

Fig. 5-5-4 Patterns of two-piece sleeves without a vent

图 5-5-4 不带袖衩的两片袖结构

Unit 6 Kimono sleeve pattern design
第六节 连身袖纸样设计

A kimono sleeve is a type of sleeve in which the sleeve is connected to the body or part of the body. The sleeve is divided into two parts, connected with the armhole of the front and back body respectively. Many ancient Chinese costumes had simple kimono sleeves. Modern sleeve-ups are common in both raglan sleeve and gusseted sleeve.

连身袖是衣袖与衣身或部分衣身相连的袖型。衣袖分成前后两部分，并分别与前、后衣身的袖窿相连。中国古代的服装很多都是简单的连身袖款式。现代的连身袖常见于插肩袖和有插角的连身袖两种。

一、Raglan sleeves / 插肩袖

A raglan sleeve is a sleeve that extends in one piece to the neckline with a seam from the armhole to the neck. The making up of the raglan sleeve is shown in Fig. 5-6-1. The structure method of this pattern is shown in Fig. 5-6-2. First, cut the front and back shoulder parts and connect them to the sleeve crown with matching points. The ease of the ordinary sleeve is concentrated in the sleeve crown, and the raglan sleeve is formed by darts instead of the ease. Fig. 5-6-2(b) shows the complete shape after the sleeve crown and the shoulder parts are connected. Of course, the position of the shoulder line can not only be limited to the change of the shoulder range, but also move over a large range of the garment body, becoming the sleeve connected with the bodice, as shown in Fig. 5-6-3.

插肩袖是一种延伸到领口的袖子，有从袖窿到颈部的接缝。插肩袖的构成如图 5-6-1 所示。这种构成方法反映到纸样上，如图 5-6-2 所示。先将前、后衣片肩部剪下来的部分以对位点对接到袖山上。普通装袖的吃势量集中在袖山头，而插肩袖是以省道代替吃势量构成曲面。图 5-6-2(b) 为袖山和衣片肩部连接后的完整形态。当然，插肩线的位置可以不只是局限在肩部范围变化，而是在衣身上的很大范围内移动，成为衣袖衣身相连。如图 5-6-3 所示。

When a raglan sleeve is used in sportswear and casual wear, it is often a looser shape; when used in suits and coats, it is a more fitted shape.

插肩袖，用在运动装、休闲装类服装中时常以宽松的形式出现，用在西服和外套中则以合体形式出现较多。

Fig. 5-6-1 Composition of raglan
sleeves
图 5-6-1 插肩袖的构成

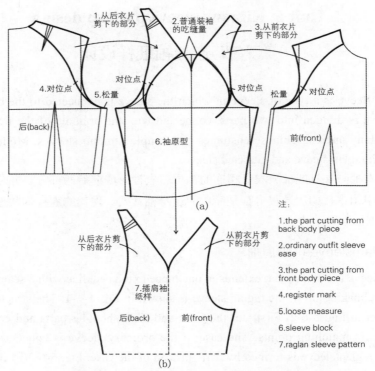

Fig. 5-6-2 Pattern composition of raglan sleeves

图 5-6-2 插肩袖的纸样构成

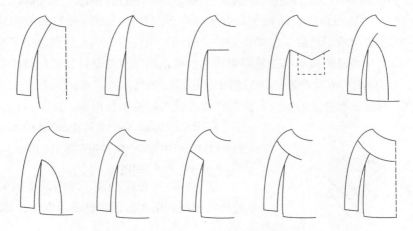

Fig. 5-6-3 Raglan position changing

图 5-6-3 插肩位置变化

1. Two-piece raglan sleeves / 两片插肩袖

For a raglan sleeve pattern, the sleeve crown is still a restricting structure factor. The angle between the sleeve midline and the shoulder line (extension line) affects how well the sleeve fits, and at the same time, the angle and the sleeve crown restrict mutually, i.e. the higher the sleeve crown height, the greater the angle, the more fitted the sleeve; the smaller the sleeve crown height, the smaller the angle, the looser the sleeve.

对于插肩袖的纸样设计，袖山高仍是制约其结构的因素。而袖中线与肩线（延长线）的夹角影响着袖的贴合度，同时它与袖山高互相制约，即：袖山越高，袖中线与肩线的夹角越大，袖子越合身；袖山越低，袖中线与肩线的夹角越小，袖子越宽松。

As an example of sleeves that are neither very fitted nor very loose, the sleeve crown height is the same as the basic sleeve crown height, i.e. AH/4 + 2.5cm. The sleeve centerline should be at an angle of 45 ° from the horizontal line. The sleeve style is as shown in Fig. 5-6-4.

例如既不十分贴体也不很宽松的袖子，其袖山高为基本袖山高，即 AH/4+2.5cm。袖中线与水平线夹角为 45°。袖子款式见图 5-6-4。

Copy the body pattern of the front and back pieces. Draw the curve of the shoulder insertion on the front and back pieces as shown in Fig. 5-6-5. Correct the shoulder line of the back piece (the shoulder ends are slightly raised by 1cm). Then draw other lines on the front and back as shown in the drawing. The following should be done when drawing: The straight line distance of CA and CB in the front piece is equal, their curve directions are opposite, their curve shape is similar, their curve length is equal; the straight line distance of DE and DF in the back piece is equal, their curve directions are opposite, their curve shape is similar, their curve length is equal. The opening width is smaller by 4 cm than the sleeve width. Draw a line from the sleeve width line to the opening line of sleeve. In order to make the sleeve fitted, the sleeve body should tilt forward, so that the front and back opening tilt forward by 3 cm separately, then draw the line smoothly again, as show in Fig. 5-6-5.

Fig. 5-6-4 Two-piece raglan sleeves
图 5-6-4 两片插肩袖的款式图

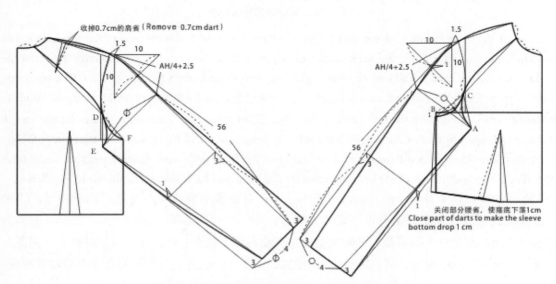

Fig. 5-6-5 Patterns of two-piece raglan sleeves
图 5-6-5 两片插肩袖的结构图

先复制前、后衣身纸样，如图 5-6-5 所示画出前、后片上插入肩部的弧线，并修正后片肩线（肩端点略抬高 1cm）。再按图中所示画出前、后片上其他各部位线条。画图时需做到：前片中 CA 与 CB 的直线距离相等，弧线方向相反、形状相似、长度相等；后片中 DE 与 DF 的直线距离相等，弧线方向相反、形状相似、长度相等。袖口比袖肥小 4cm 左右，从袖宽线向袖口画线。为使袖子合体，袖身需往前倾斜，因此前、后袖口处分别往前倾斜 3cm，然后重新画顺弧线即可，见图 5-6-5。

2. Loose raglan sleeves / 宽松插肩袖

Loose bodice should match loose sleeves. The sleeve crown height of loose raglan sleeves should be smaller than that of general sleeves. This loose raglan sleeve is 57 cm long and its sleeve opening width is 15 cm, and the top-point of the sleeve crown is 0.5 cm away from the shoulder point, and the front and back sleeve tilt degrees are shown in Fig. 5-6-6.

宽松的衣身应该匹配宽松的袖子。宽松插肩袖的袖山高应该比一般袖型的袖山高小。这款宽松插肩袖的袖长 57cm，袖口宽 15cm。袖山高点离开肩点 0.5cm，前、后袖斜度见图 5-6-6 所示。

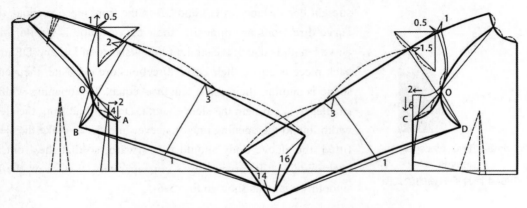

Fig. 5-6-6 Patterns of loose raglan sleeves
图 5-6-6 宽松插肩袖结构图

Lower the front armhole 4 cm and lower the back armhole 6 cm, reconnect the armhole curve smoothly, and make the front and back sideseams equal. Divide the curve below the intersection of the armhole curve and bust width line or back width line into three equal sections, and the first trisection point is point O. Set the sleeve crown height as AH/4-2.5cm, draw a sleeve width line according to the sleeve crown height on the back piece, find point B on the sleeve width line, and make the length of curve OB equal to that of curve OA. Similarly, the length of curve OD equal to that of curve OC on the front piece. There is a difference of 2 cm between the front and back sleeve openings. Draw lines from the sleeve width lines to the opening lines on the front and back pieces, as shown in Fig. 5-6-6.

将后片窿深降低 4cm，前片窿深降低 6cm，重新连顺袖窿弧线并使前后侧缝相等。将袖窿弧线与胸宽或背宽线的交点以下弧线三等分，第一个等分点即为袖标点 O。设定袖山高为 AH/4-2.5cm，在后片上根据袖山高画袖宽线，在袖宽线上找到 B 点，使 OB 与 OA 相等。同理，在前片上使 OD 与 OC 相等。前、后袖口大相差 2cm。在前、后片上分别从袖宽线向袖口线画线，见图 5-6-6。

二、Gusseted sleeves / 插角袖

Design the gusseted sleeve based on the bodice prototype. Set the sleeve length as 56 cm and the sleeve opening width as 14 cm. Align the back waist line with the position on half of the front "falling amount", and scoop the other half at the bottom of the front armhole, link the armhole, and straighten side seams of the front and back bodice block. Take point D and point C on the front and rear side seams respectively, so that the distances between them to the front and back hemlines are both 10 cm.

以原型上衣为基础来设计插角袖。设插角袖的袖长为 56cm，袖口宽为 14cm。将原型上衣的后片腰线与前片下落量的一半对齐，前袖窿下挖前片下落量的另一半，连顺袖窿，并将前、后片的侧缝线拉直。分别在前后侧缝上取 D 点和 C 点，使它们到前、后底摆线的距离均为 10cm。

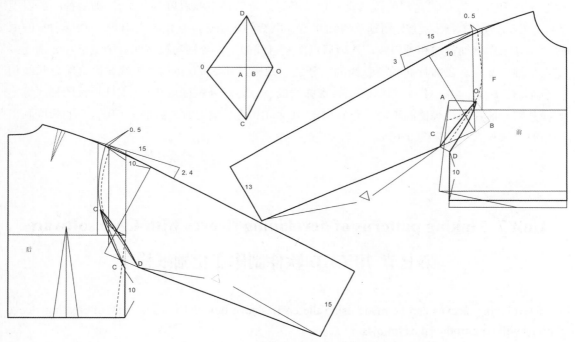

Fig. 5-6-7 Patterns of gusseted sleeves

图 5-6-7 插角袖结构图

In general gusseted sleeves are included in loose kimono sleeves. If the shoulder line and the midline of loose kimono sleeve are not connected in a straight line, the position of the sleeve midline should be drawn according to the angle between the extended shoulder line and the sleeve midline. However, the method of drawing a line by measuring the angle with a protractor is not applicable in garment drawing and has gradually been replaced by a simpler method. For example, you can use a slope of 15 : x to represent the drawing of a diagonal line (15 and X are the lengths of the two right-angled sides of a right triangle). If the front sleeve midline slope is 15 : x, the back sleeve midline slope is 15 : 0.8x. On the front and back pieces, extend the shoulder line by 0.5 cm to the new shoulder point, and the sleeve crown height is 10 cm. Make the sleeve width line and the sleeve opening line according to the pattern design method of the raglan sleeve, make a vertical line down from the new shoulder point, and have an intersection with the sleeve width line respectively, and connect the intersection point and the sleeve opening point respectively. Mark point O at a third of the new armhole curve on the front piece. Make a horizontal line from point O to intersect with the extension line of side seam at point A. Connect point O with points C and D to make OC = OD. And point O is also marked at one third of the back armhole curve. Under the condition that the sleeve bottom seam and side seam of the back piece are equal to that of the front piece respectively, OC and OD are drawn from point O. Draw the shape and size of the insert angle according to the lengths of OD and OC on the front and back sleeves, as shown in Fig. 5-6-7.

插角袖一般属于宽松连身袖。当宽松连身袖的肩线与袖中线不是连成一条直线时，需要根据肩线延长线与袖中线的夹角画出袖中线的位置。但用量角器测量角度来画一条线的方法在服装制图中不适用，已逐步被更简单的方法取代。比如，可以用斜度为 15:x 来表示一条斜线的画法（15 和 X 是直角三角形的两条直角边长度）。若前袖中线斜度为 15:x，则后袖中线斜度为 15:0.8x。在前、后片上分别延长肩线 0.5cm 至新的肩端点，袖山高为 10cm，按插肩袖的纸样设计方法作出袖宽线和袖口线，从新肩点作竖直线向下，分别与袖宽线有个交点，分别连接交点与袖口宽点。在前片新袖窿弧线的三分之一处标点 O，从 O 点作水平线与侧缝的延长线相交于 A，连接 OC、OD，使 OC=OD。在后袖窿弧线的三分之一处也标点 O。在保证后片的袖底缝、侧缝分别与前片的相等情况下，从点 O 引出 OC 和 OD 线。根据前、后袖上 OD、OC 的长度画出插角的形状和大小，见图 5-6-7。

Unit 7 Making patterns of developing sleeves with CAD software
第七节　用 CAD 软件制作变化袖纸样

Developing sleeves can be made using the basic sleeve block. The usual way of creating sleeve patterns will be introduced as follows.

变化的袖片制图可在原型袖片基础上进行。下面介绍几种常见袖型的制作。

一、Puff sleeves / 泡泡袖

As shown in Fig. 5-7-1, it is a style of typical puff sleeves. Some pleat volumes are added in the crown of the basic sleeve. The steps are as follows:

如图 5-7-1 所示为典型的泡泡袖款式。袖山处在基本袖片基础上增加了抽褶量。步骤如下：

① Take out the basic sleeve pattern – draw cutting lines, as shown in Fig. 5-7-2.

取袖子基本形样片—作剪切线，如图 5-7-2 所示。

② Select "handle pattern – open three-dimensional – open both sides", select "datum line" – select "cutting line" in order – right-click to end – input data – Determine. Get the new sleeve crown as shown in Fig. 5-7-3.

选择"样片处理—立体展开—两侧展开"，选择"展开基准线"——再依次选择"分割线"——右击结束——输入相关数据——确定。得如图 5-7-3 所示袖山。

Fig. 5-7-1 The style of
typical puff sleeves
图 5-7-1 典型的泡泡袖款式

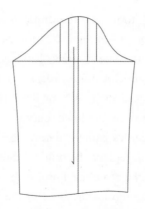

Fig. 5-7-2 Make cutting lines
图 5-7-2 作剪切线

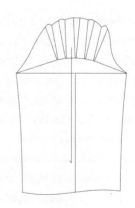

Fig. 5-7-3 Open sleeve crown
图 5-7-3 袖山展开

二、lantern sleeve / 灯笼袖

It is a lantern sleeve style with dividing lines at the bottom cuff and adding ease at the dividing lines, as shown in Fig. 5-7-4. You can use the tools of "handle pattern – cut and move pattern" and "cutting pattern". The steps are as follows:

如图 5-7-4 所示为灯笼袖款式，袖口有分割线，分割线处要放松量。制作时需要利用 "样片处理—样片剪开移动" 和 "分割样片" 工具。步骤如下：

① Copy the sleeve prototype and shorten the sleeve length to the elbow line as shown in Fig. 5-7-5. The sleeve crown height is increased by 1 cm and the sleeve width is narrowed. Make several vertical cutting lines from the sleeve crown curve to the sleeve elbow line, and make a transverse dividing line 7 cm from the sleeve elbow line. The patterns of the upper and lower parts splitted by the dividing line are taken out separately, as shown in Fig. 5-7-5.

复制袖子原型，如图 5-7-5 所示将袖长改短至袖肘线。袖山高调高 1cm，袖宽变窄。从袖山弧线作几条垂直切割线至袖肘线，并在距袖肘线 7cm 处作横向分割线。将分割线上下二部分分别取样片，如图 5-7-5 所示。

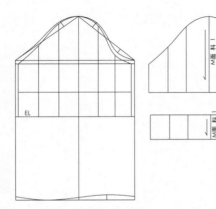

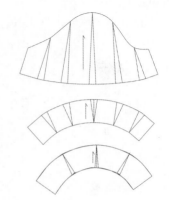

Fig. 5-7-4 The lantern
sleeve style
图 5-7-4 灯笼袖款式

Fig. 5-7-5 Draw a transverse cutting
line and split the sleeve
图 5-7-5 作剪切线并上下分割袖片

Fig. 5-7-6 Cut and add pleat values
图 5-7-6 剪开加入松量

② Select "Cut and move pattern" tool – on the upper piece, click the cutting lines (in anti-clockwise order) – right-click, and input values: input "2" for the lower end of each cutting line, and input "0" for the upper end of each cutting line – select "Round" for both ends – Determine – on the lower piece, click the cutting lines (in anticlockwise order) – right-click, and input values: input "2" for the upper end of each cutting line, and input "0" for the lower end of each cutting line – select "Round" for both ends – Determine – redraw the lower curve of the upper piece and the upper curve of the lower piece, and make the two curves equal – redraw the lower curve of the lower piece, and using the "Replacement net outline" tool, replace the original three curves – on the lower piece, using "Delete internal elements" tool, delete the structure line inside – draw three new cutting line – use "Cut and move pattern" tools shortening the lower curve of the lower piece by total 3 cm (inputing negative value can shorten the line, so input "-1" for the lower end of each cutting line) – finally, using the "Move" tool, place the patterns straight and change the yarn direction. As shown in Fig. 5-7-6.

选择"样片剪开移动"工具—在上片上点击剪切线（按逆时针顺序）—右击，输入数值：每条剪切线的下端加放 2cm，上端（袖山弧线处）加放 0—选"圆顺"工具对上下两端进行圆顺处理—确定—在下片上点击剪切线（按逆时针顺序）—右击，输入数值：每条剪切线的上端处加放 2cm，下端处加放 0—上下端都选"圆顺"—确定—重新画顺上片的下口线和下片的上口线，并使这两条弧线长相等—重新画顺下片的下口线，用"样片换净边"工具，替换原来的 3 条弧线—在下片上用"删除样片内部元素"工具，删掉结构内线—重新画 3 条切割线—用"样片剪开移动"工具，使下片上的下口线共缩短 3cm（输入负数可使之变短，因此每条剪切线下端输入"-1"）—最后，用"移动"工具，把样片都放正并修改纱向。如图 5-7-6 所示。

三、Petal sleeves / 花瓣袖

The petal sleeve style is shown in Fig. 5-7-7. The steps are as follows:
如图 5-7-7 所示为花瓣袖款式。步骤如下：

① Take the basic sleeve – draw two cutting curves, as shown in Fig. 5-7-8.
取袖基本形—作二条分割弧线，如图 5-7-8 所示。

② Use the tool of "take out pattern" – take out the front and back pieces of the sleeve, as shown in Fig. 5-7-9.
利用"样片取出"工具，分别将前、后袖片取出即可，如图 5-7-9 所示。

Fig. 5-7-7 The petal sleeve style
图 5-7-7 花瓣袖款式

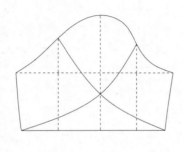

Fig. 5-7-8 Draw cutting curves
图 5-7-8 作分割弧线

后袖　前袖

Fig. 5-7-9 The front and back pieces of petal sleeves
图 5-7-9 花瓣袖前、后片

Exercise / 练习

1. Design 3 or 4 sleeveless styles and make their patterns.

设计 3 或 4 款无袖款式并制作它们的样板。

2. Make patterns of one-piece fitted sleeves.

制作合体一片袖样板。

3. Make patterns of two-piece fitted sleeves.

制作合体两片袖样板。

4. Design 3 or 4 styles of developing sleeves and make their patterns.

设计 3 或 4 款变化袖型并制作它们的样板。

Chapter 6 Comprehensive training
第六章 综合训练

Unit 1 Skirt pattern design
第一节 裙子纸样设计

A skirt's basic shape is simple. When a person is stood up straight, it looks like a cylindrical structure that wraps around the waist, stomach, hips and legs.

裙子的基本形状比较简单。它是在人体直立姿态下围裹人体腰部、腹部、臀部、下肢一周所形成的筒状结构。

This unit is mainly about the pattern design of basic skirts and how to make the developing patterns of A-line skirt, trumpet shape skirt and other developing skirts according to the pattern of the basic skirt.

这一节内容主要讲解基本裙的纸样设计，并在基本裙纸样基础上变化出 A 型裙、喇叭裙及其他各种变化款裙的纸样。

一、The basic skirt / 基本裙

According to the style characteristics shown in Fig. 6-1-1, the skirt specifications are set as follows: size designation is 160/68A; skirt length is 60 cm; waist is 68cm (net) plus 0–2 cm (basic ease); hip is 90 cm (net) plus 4 cm (basic ease); waist-length（from waist line to hip line）is 18 cm; waistband height is 4 cm. A zipper is at the top of the back centerline, and a vent is at the bottom.

根据款式如图 6-1-1 所示特征，该基本裙的成衣规格设定如下：号型 160/68A，裙长为 60cm，腰围 68cm（净）+0 ～ 2cm（基本松量），臀围 90cm（净）+4cm（基本松量），腰长（从腰围线到臀围线）18cm，腰头宽 4cm。后中线上装拉链下开衩。

The pattern making method of a basic skirt is as shown in Fig. 6-1-2. The ease has

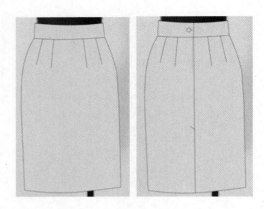

Fig. 6-1-1 A basic skirt shape
图 6-1-1 基本裙型

been added to the waist and hips in these formulas. The back waist line is 1 cm lower than the center of the front waist line. The front and back waistlines uprise 0.7–1 cm separately at the side seam. There are two separate darts on the front and back pieces. Each dart size is one-third of the difference in amount between the waist and the hip.

　　基本裙的纸样制作方法如图 6-1-2 所示。公式中的腰围和臀围都已加上松量。后腰围线在后中心线处低落 1cm，前、后腰围线在侧缝处分别起翘 0.7 ～ 1cm。前、后片各有两个省。每个省的大小分别为前后片的臀腰差量的 1/3。

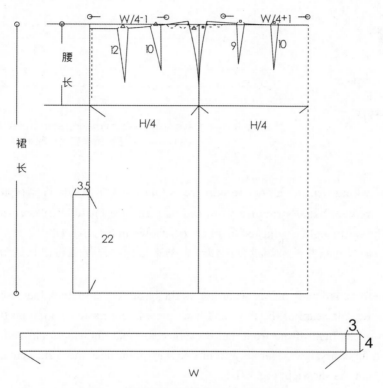

Fig. 6-1-2 pattern of a basic skirt
图 6-1-2 基本裙纸样

二、Developing skirts / 变化款裙

1. A-line skirts / A 字裙

According to the style characteristics shown in Fig. 6-1-3, the skirt specifications are set as follows: size designation is 160/68A; skirt length is 40cm; waist is 68 cm (net) plus 0–2 cm (basic ease); hip is 90 cm (net) plus 4 cm (basic ease); waist-length is 18 cm; waistband height is 4 cm. The actual finished A-line skirt has more than 4 cm ease of hip. (The style of the back piece of the A-line skirt is similar to the front, and the zipper is at right side.)

　　根据款式如图 6-1-3 所示特征，该基本裙的成衣规格设定如下：号型 160/68A，裙长为 40cm，腰围 68cm（净）+0 ～ 2cm（基本松量），臀围 90cm（净）+4cm（基本松量），腰长 18cm，腰头宽 4cm。最终实际的成品 A 字裙的臀围松量大于 4cm。（A 字裙的后片款式与前片相似，

Fig. 6–1–3 The style of
A–line skirts

图 6–1–3 A 字裙款式

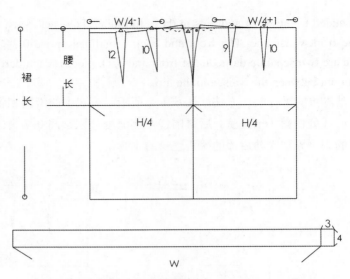

Fig. 6–1–4 Developing pattern making step 1 of A–line skirts
图 6–1–4 A 字裙纸样变化步骤 1

拉链装在右侧）。

The pattern-making steps for an A-line skirt are as follows/ A 字裙的纸样制作步骤如下：

① Firstly，make a basic structure from the pattern of a basic skirt according to the skirt specifications. The skirt length is adjusted to 40 cm (as shown in Fig. 6-1-4).

首先根据规格尺寸按照基本裙纸样制作一个基本结构，只需要将裙长调节为 40cm。如图 6-1-4 所示。

② Copy the front and back pieces according to the basic structure, then increase the hem by 3–4 cm separately at the side seam of the front and back pieces, and make cocking up 0.8 cm separately at the side seam of the hem of the front and back pieces. Make the side seam and hem curve lines smooth. Draw an auxiliary line separately from one dart point near the front centerline or the back centerline to the hem. As shown in Fig. 6-1-5.

复制前、后裙片基本结构，在前、后裙片的侧缝下摆处分别放出 3 ～ 4cm 以便增大下摆，

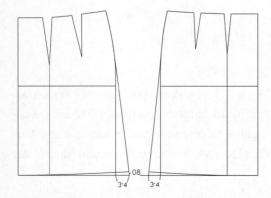

Fig. 6–1–5 Developing pattern making step 2 of
A–line skirts

图 6–1–5 A 字裙纸样变化步骤 2

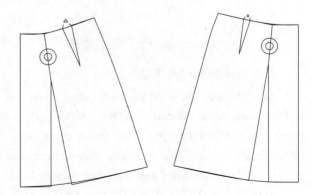

Fig. 6–1–6 Developing pattern making step 3 of A–line skirts

图 6–1–6 A 字裙纸样变化步骤 3

前、后片下摆侧缝处分别起翘 0.8cm。修顺侧缝线和下摆线。从靠近前、后中心线的省尖处分别画一条辅助线至下摆。如图 6-1-5 所示。

③ Cut the front and back blocks along the auxiliary line separately to the dart point, and then fold the darts of the front and back pattern pieces to transfer the dart value to the hem as hem ease value. As shown in Fig. 6-1-6.

沿着辅助线分别剪开前、后样片至省尖点，然后分别折叠前、后片的省道使省道量转移到下摆处，作为下摆松量。如图 6-1-6 所示。

④ The waist and the hem lines must be smooth curves. Adjust the positions of the other darts to the middle of the front and back waist lines separately. As shown in Fig. 6-1-7.

修正腰围线和下摆线使之光滑圆顺。分别调整前、后裙片上剩下的腰省位置到腰围线的中间位置。如图 6-1-7 所示。

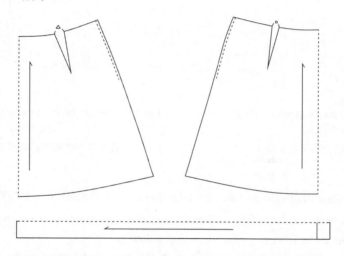

Fig. 6-1-7 Developing pattern making step 4 of A-line skirts

图 6-1-7 A 字裙纸样变化步骤 4

2. Flare skirts / 喇叭裙

According to the style characteristics shown in Fig. 6-1-8, the skirt specifications are set as follows: size designation is 160/68A; skirt length is 70 cm; waist is 68 cm (net) plus 0–2 cm (basic ease); waist-length is 18 cm and waistband height is 4 cm. (The back piece style of flare skirt is similar to the front piece style, and the zipper is at the right side.)

根据款式如图 6-1-8 所示特征，该基本裙的成衣规格设定如下：号型 160/68A，裙长为 70cm，腰围 68cm（净）+(0～2cm)（基本松量），腰长 18cm，腰头宽 4cm。（喇叭裙的后片款式与前片相似，拉链装在右侧。）

Fig. 6-1-8 A flare skirt style

图 6-1-8 喇叭裙款式

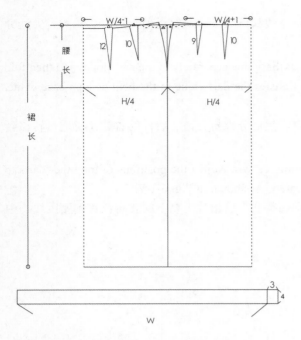

Fig. 6-1-9 Developing pattern making step 1 of
flare skirts

图 6-1-9 喇叭裙纸样变化步骤 1

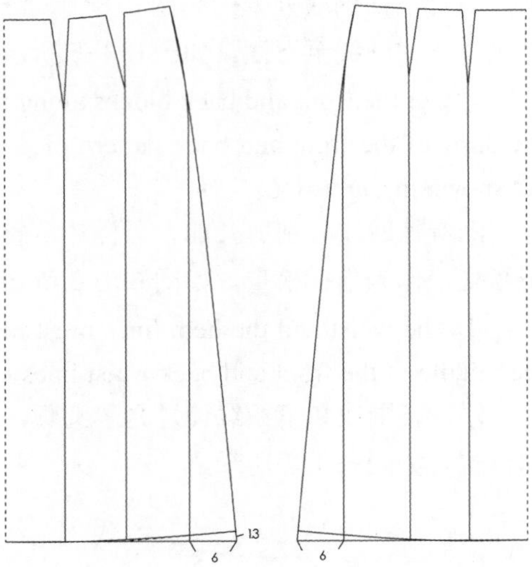

Fig. 6-1-10 Developing pattern making step 2 of
flare skirts

图 6-1-10 喇叭裙纸样变化步骤 2

The steps for making flare skirt patterns are as follows/ 喇叭裙的纸样制作步骤如下：

① First, make a basic structure from the basic skirt pattern according the skirt specifications, and the skirt length is adjusted to 70 cm. As shown in Fig. 6-1-9.

首先根据规格尺寸按照基本裙纸样制作一个基本结构，并将裙长调节为 70cm。如图 6-1-9 所示。

② Copy the front and back pattern pieces according to the basic structure, then increase the hem by 6 cm at the side seams of the front and back pattern pieces. And make cocking up 1.3 cm separately at the side seam of the front and back hem. The side seam and the hem lines must be smooth curves. Draw two auxiliary lines separately from the dart points to the hem. As shown in Fig. 6-1-10.

复制前、后裙片基本结构，然后在前、后裙片的侧缝下摆处放出 6cm 以便增大下摆。在前、后 p 下摆的侧缝处分别起翘 1.3cm。修圆顺侧缝线和下摆线。从省尖处分别画两条辅助线至下摆。如图 6-1-10 所示。

③ Cut the front and back blocks along the auxiliary lines separately to the dart points, then on the front and back pattern pieces fold the darts separately to transfer the dart value to the hem as hem ease value. As shown in Fig. 6-1-11.

分别沿着辅助线剪开前、后样片至省尖点，然后在前后片上分别折叠省道使省道量转移到下摆处，作为下摆松量。如图 6-1-11 所示。

④ The waist and hem line must be smooth curves. As shown in Fig. 6-1-12.

修正腰围线和下摆线使之光滑圆顺。如图 6-1-12 所示。

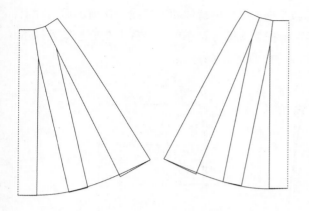

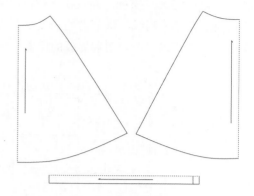

Fig. 6-1-11 Developing pattern making step 3 of flare skirts

图 6-1-11 喇叭裙纸样变化步骤 3

Fig. 6-1-12 Developing pattern making step 4 of flare skirts

图 6-1-12 喇叭裙纸样变化步骤 4

3. Skirts with yoke style lines and pleats / 育克分割褶裥裙

According to the style characteristics shown in Fig. 6-1-13, the skirt specifications are set as follows: size designation is 160/68A; skirt length is 65 cm; waist is 68 cm (net) plus 0-2 cm (basic ease); hip is 90 cm (net) plus 4 cm (basic ease); waist-length is 18 cm; waistband height is 4 cm. The final finished skirt with yoke style lines and pleats has more than 4 cm ease of hip.（The back piece style of the skirt with yoke style lines and pleats is similar to the front, and the zipper is at right side.）

根据如图 6-1-13 所示款式特征，该基本裙的成衣规格设定如下：号型 160/68A，裙长 =65cm，腰围 =68cm（净）+0 ~ 2cm（基本松量），臀围 =90cm（净）+4cm（基本松量），腰长 =18cm，腰头宽 =4cm。最终育克分割褶裥裙成品的臀围松量大于 4cm。（育克分割褶裥裙的后片款式与前片相似；拉链装在右侧。）

Fig. 6-1-13 A skirt with yoke style lines and pleats

图 6-1-13 育克分割褶裥裙

The pattern-making steps for the skirt with yoke style lines and pleats are as follows/ 育克分割褶裥裙的纸样制作步骤如下：

① First, make a basic structure from the basic skirt pattern according the skirt specifications, and the skirt length needs to be adjusted to 65 cm. Increase the hem by 4 cm separately at the side seam of the front and back pattern pieces, and also make cocking up 1 cm at the side seam of the hem. The side seam and hem line must be smooth curves. Draw the yoke style lines separately on the front and back pattern pieces. Adjust the dart length to lengthen it to the yoke style line. Draw two auxiliary lines to the hem separately as the pleat positions. As shown in Fig. 6-1-14.

首先根据规格尺寸按照基本裙纸样制作一个基本结构，只需要将裙长调节为 65cm。在前、后裙片的侧缝下摆处放出 4cm 以便增大下摆量，并且在下摆侧缝处起翘 1cm。修顺侧缝线和下摆线。在前、后裙片上分别画出育克分割线，调节省的长度至分割线。根据款式图分别在前、后片上从分割线处往下摆画两条辅助线，作为褶裥加放的位置。如图 6-1-14 所示。

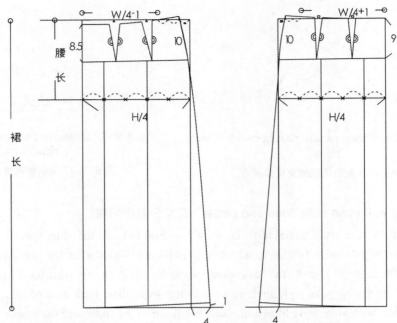

Fig. 6-1-14 Developing pattern making step 1 of the skirt with yoke style lines and pleats

图 6-1-14 育克分割褶裥裙的纸样制作步骤 1

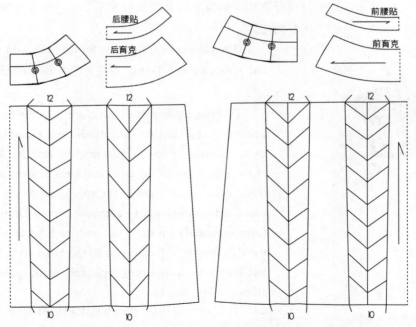

Fig. 6-1-15 Developing pattern making step 2 of the skirt with yoke style lines and pleats

图 6-1-15 育克分割褶裥裙的纸样制作步骤 2

② Cut along the yoke style lines separately on the front and back pattern pieces to create two parts. In the upper part, fold the darts to hide them in the style line of yoke and draw the front and the back waist welts. In the lower part, cut along two auxiliary lines from the yoke style line to the hem separately on the front and back pattern pieces, and making a 12 cm spread horizontally. Adjust each pleat so that its top is 12 cm and its bottom is 10 cm (as shown in Fig. 6-1-15).

分别沿着前、后裙片的分割线剪开，以生成为上、下两部分。在上片部分，折叠省道使之转移到分割线处隐藏，画出前、后裙片的腰贴。在下片部分，分别在前、后裙片上沿着两条辅助线从育克分割线处剪到下摆，水平拉开 12cm。调整每个褶裥量，使其上端为 12cm、下端为 10cm（如图 6-1-15 所示）。

Unit 2 Shirt pattern design
第二节 衬衫纸样设计

The bust ease in common clothing is as follows: on the basis of a net bust, add 4–6 cm for bust of a tight dress or QiPao, add 6–8 cm for bust of a fitted shirt, add 8–10 cm for bust of a fitted coat or jacket, add 12–14 cm for bust of a Spring/Autumn fitted suit, add 16–18 cm for bust of a fitted trench coat and other general coat, and add 20–24 cm for bust of a loose trench coat or winter coat.

常见服装品类的胸围放松量如下：在净胸围的基础上，紧身连衣裙或旗袍类胸围加放 4 ～ 6cm，合体衬衫胸围加放 6 ～ 8cm，合体上衣或夹克胸围加放 8 ～ 10cm，合体春秋套装胸围加放 12 ～ 14cm，合体风衣或一般外套大衣胸围加放 16 ～ 18cm，宽松风衣或冬季大衣等胸围加放 20 ～ 24cm。

一、The blouse with gathers on placket / 胸前门襟有细褶女衬衫

Based on the style characteristic, set the shirt specifications as follows: size designation is 160/84A; shirt back length is 58 cm; bust is 94 cm; waist is 80 cm; shoulder width is 38 cm; sleeve length is 58 cm; cuff is 20 cm. As shown in Fig. 6-2-1.

根据款式特征，该衬衫的成衣规格设定如下：号型 160/84A，后衣长 58cm，胸围 94cm，腰围 80cm，肩宽 38cm，袖长 58cm，袖口 20cm。如图 6-2-1 所示。

First copy the body prototype of the front and back pattern pieces, lengthen the back piece by 20 cm, and so do the front piece. On the back piece, slightly widen the back neckline, correct the back shoulder line and back armhole curve, and draw the back side-seam curve and the hem curve, and make the waist dart. On the front piece, a part of the bust dart value of the prototype front piece was transferred to the front centerline for making chest slope, and the remaining was used for the side seam dart. Widen the front neckline and correct the front shoulder line and the front armhole curve so that the front shoulder line is 0.5 cm shorter than the back shoulder line. Draw the front side seam curve and the hem curve; Take a 1.8 cm wide overlap to make the placket, and draw the placket structure and buttons position. As shown in Fig. 6-2-2.

先复制前、后片衣身原型，在后片上加长 20cm，在前片上也加长 20cm。在后片上，略开大后领口，修正后肩线、后袖窿弧线，画顺后侧缝弧线及下摆弧线，作出腰省。在前片上，将原型前片的胸省转移一部分到前中心线处作为撇胸，剩余部分作为侧缝省。开大前领口，修正前肩线和前袖窿弧线，使前肩线长比后肩线长短 0.5cm。画顺前侧缝弧线及下摆弧线；取搭门量 1.8cm，画出门襟止口结构和纽位。如图 6-2-2 所示。

On the front piece, transfer the side seam dart to the front placket, and the volume transferred is taken as the gathers. Draw several auxiliary lines, cut and open them to increase the gathers. As shown in Fig. 6-2-2. Sleeve structure: First measure the front armhole curve length (Front AH) and the back armhole curve length (Back AH). Sleeve midline length is sleeve length (58 cm) minus cuff width (6

Fig.6-2-1 The blouse style with gathers on placket

图 6-2-1 胸前褶皱女衬衫款式

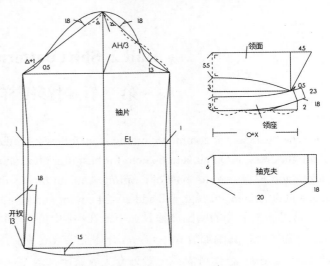

Fig. 6-2-3 Pattern 2 of the blouse with gathers on placket

图 6-2-3 胸前褶皱女衬衫结构图二

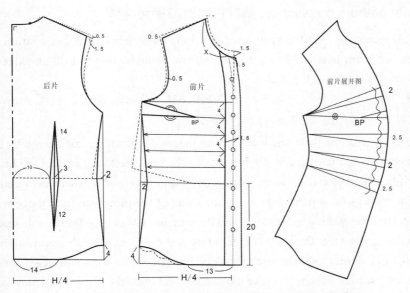

Fig.6-2-2 Pattern 1 of the blouse with gathers on placket

图 6-2-2 胸前褶皱女衬衫结构图一

cm) equal to 52 cm; Sleeve crown height is 1/3 of the total AH; The length of the front sleeve crown slant line is the front AH, and the length of the back sleeve crown slant line is the back AH plus 0.5 cm; From the top, the position taking 1/2 of sleeve length plus 2.5 cm is the sleeve elbow line, and other sleeve drawing details and sleeve cuff are drawn as shown in Fig. 6-2-3. The collar structure is also shown in Fig. 6-2-3.

在前片上将侧缝省转移到门襟处，转移的量作为碎褶量。再作几条辅助线并剪开、拉开以加大褶量，见图 6-2-2。袖子结构：先分别测量前袖窿弧线长（前 AH）、后袖窿弧线长（后 AH）。袖中线长 = 袖长（58cm）− 袖克夫宽（6cm）=52cm；袖山高为总 AH/3；前袖山斜线长为前 AH，后袖山斜线长为后 AH+0.5cm；从顶点取袖长 /2+2.5cm 处为袖肘线，其余袖子制图细节和袖克夫绘制如图 6-2-3 所示。领子结构也如图 6-2-3 所示。

二、The shirt with short sleeves and stand collars / 立领短袖衬衫

Based on the style characteristics, set the shirt specifications as follows: size designation is 160/84A; shirt back length is 80 cm; bust is 90 cm; shoulder width is 39 cm; sleeve length is 15 cm; cuff width is 30 cm. See Fig. 6-2-4.

根据款式特征，该衬衫的成衣规格设定如下：号型 160/84A，后衣长 80cm，胸围 90cm，肩宽 39cm，袖长 15cm，袖口 30cm。见图 6-2-4。

On the front bodice block, transfer a part of the bust dart value to make the waist line level. Transfer some of the dart value to the front centerline for making a chest slope, and the remaining of

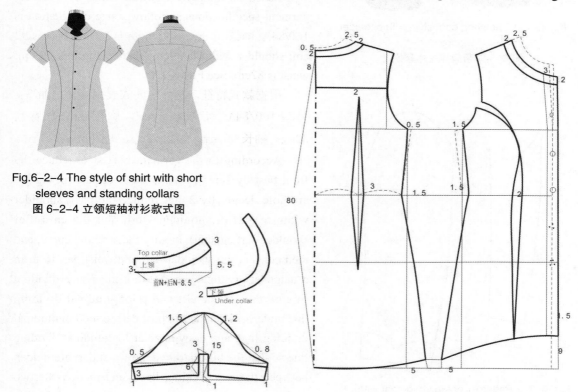

Fig.6-2-4 The style of shirt with short sleeves and standing collars
图 6-2-4 立领短袖衬衫款式图

Fig. 6-2-5 Patterns of the shirt with short sleeves and standing collars
图 6-2-5 立领短袖衬衫结构图

that to the armhole, and connect the armhole dart with the waist dart forming the dividing line. Adjust the back shoulder width to "shoulder width /2+0.5cm=20cm" according to the garment so that the front shoulder line length is 0.5 cm smaller than the back's. Adjust the bust of the front or back bodice according to the garment: narrow the front bust by 1.5 cm and the back bust by 0.5 cm. Set the yoke width at the back shoulder as 8 cm. The structure of other parts is shown in Fig. 6-2-5.

转移一部分前片原型上的胸省以使腰围线水平。转省时，把一部分量转到前中线作为撇胸（1cm），剩下的量转到袖窿上，与腰省连成分割线。根据成衣肩宽修正后片肩宽，使其为肩宽 /2+0.5cm=20cm，因此前肩线长度比后肩线短 0.5cm。根据成衣胸围大小修改前、后片胸围：前片胸围缩小 1.5cm，后片胸围缩小 0.5cm。后肩部育克宽设定为 8cm。其他部件结构如图 6-2-5 所示。

Fig. 6-3-1 The short coat style with shoulder loops

图 6-3-1 带肩袢短外套款式

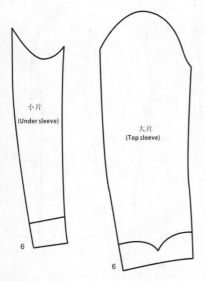

Fig. 6-3-2 Pattern 1 of the short coat with shoulder loops

图 6-3-2 带肩袢短外套结构图一

Unit 3 Outerwear pattern design

第三节 外套纸样设计

一、The short coat with shoulder loops / 带肩袢短外套

Based on the style characteristics, set the coat garment specifications as follows: size designation is 160/84A; back of the coat length is 54 cm; bust is 92 cm; shoulder width is 39cm; sleeve length is 59 cm; waist is 82cm. See Fig. 6-3-1.

根据款式特征，该外套的成衣规格设定如下：号型 160/84A，后衣长 54cm，胸围 92cm，肩宽 39cm，袖长 59 cm，腰围 82cm。见图 6-3-1。

According to the garment bust size, reduce the front bust by 1cm. To make the coat pattern, dig the armhole deeper by 2 cm, reduce the back shoulder width to half of shoulder width plus 0.5 cm (after shoulder dart closing), modify the armhole curve, and then connect the shoulder dart and waist dart to form a diagonal dividing line. Make a transverse dividing line on the front chest and a longitudinal dividing line under it. Transfer the bust dart to the longitudinal dividing line, and draw another longitudinal dividing line so that the bust dart and the waist dart are hidden between the two dividing lines, and modify the two dividing lines to be smooth. The structure of other parts is shown in Fig. 6-3-2 and Fig. 6-3-3.

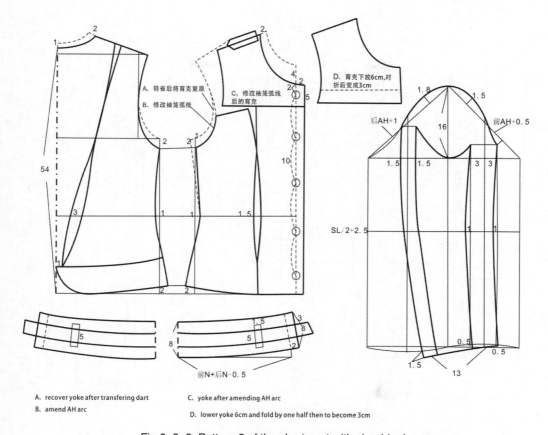

A. recover yoke after transfering dart
B. amend AH arc

C. yoke after amending AH arc
D. lower yoke 6cm and fold by one half then to become 3cm

Fig.6-3-3 Pattern 2 of the short coat with shoulder loops
图 6-3-3 带肩袢短外套结构图二

根据成衣胸围大小，将原型的前胸围减小 1cm。制作外套纸样，把袖窿挖深 2cm，后肩宽缩小至肩宽 /2+0.5cm（收肩省后），修改后袖窿弧线，再将肩省与腰省连成一条斜向分割线。在前片胸部作横向分割线，在横向分割线下作纵向分割线。转移胸省至纵向分割线处，画出另一条纵向分割线，使胸省和腰省隐藏在两条分割线之间，并修顺两条分割线。其他部位结构的作法如图 6-3-2、图 6-3-3 所示。

二、The long coat with lapels / 翻驳领长外套

This coat is a fitted structure with eight pieces, wide lapel and double-breasted narrow button stand. The front piece has a knife-back seam and the front hem is beveled. There is a transverse dividing line at the back; Under the transverse division line ,there is a crack line in the middle and there are the longitudinal division lines; There is also a transverse segment at the back of the waist. The sleeves are two-piece fitted sleeves with sleeve vents, as shown in Fig. 6-3-4.

此款外套为八片合体结构，宽驳领，双排扣窄搭

Fig.6-3-4 The long coat style with lapels
图 6-3-4 翻驳领长外套款式图

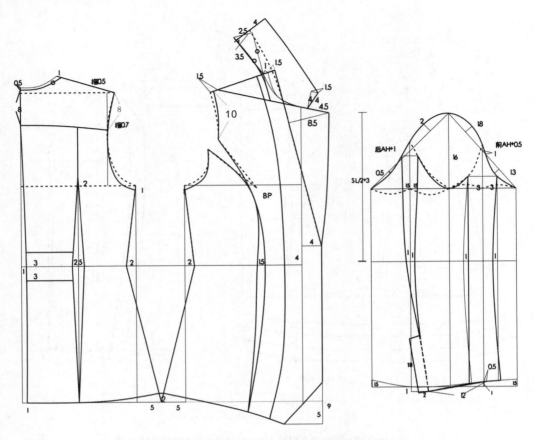

Fig. 6-3-5 Patterns of the long coat with lapels

图 6-3-5 翻驳领长外套结构图

门。前片有刀背缝线，前下摆呈斜角形。后肩背处有横向分割线；横向分割线下后中线破缝，另有纵向分割线；后腰处又有横向分割片。袖子为带袖衩的两片合体袖，如图 6-3-4 所示。

Based on the style characteristics, set the coat garment specifications as follows: size designation is 160/84A; back coat length is 72 cm; bust is 94 cm; shoulder width is 39 cm; sleeve length is 58 cm, and waist is 76cm.

根据款式特征，该外套的成衣规格设定如下：号型 160/84A，后衣长 72cm，胸围 94cm，肩宽 39cm，袖长 58cm，腰围 76cm。

Transfer the bust dart of the front bodice block to the armhole to make the waist line level, and align the front waist line with the back waist line horizontally. Adjust the shoulder width of the back piece to shoulder width/2+0.5cm=20cm, based on the shoulder width of the garment. Adjust the length of the front shoulder line to make it equal to the length of the back shoulder line minus 0.5 cm. In order to increase the size of the armhole, the front and back armhole bottoms drop 1 cm respectively and modify the new armhole curve. The patterns of other parts are shown in Fig. 6-3-5.

将前片原型胸省转移至袖窿以使腰围水平，并将前片腰围线与后片腰围线水平对齐。按成衣肩宽，将后片肩宽调整至为肩宽 /2+0.5cm=20cm。调整前肩线长，使之等于后肩线长 -0.5cm。为加大袖窿，前、后窿底各下降 1cm，修改袖窿弧线。其他部件结构如图 6-3-5 所示。

Unit 4 Making patterns with CAD software
第四节　用 CAD 软件制作纸样

The pattern-making process will be introduced mainly through the straight skirt, blouse and coat in this unit. Other styles can be created according to the same process.

本节主要介绍直裙、衬衫和外套的样板制作过程。其他款式可依此类推。

一、Straight skirts / 直裙

The specifications of the straight skirt: skirt length is 54 cm; waist is 70 cm; hip is 94 cm; waistband width is 4 cm. The patterns are as shown in Fig. 6-4-1. There are two darts in each pattern of the back and front pieces, and a vent at the lower back centerline and a zipper at the back waist.

直裙规格：裙长 54cm、腰围 68cm、臀围 90cm、腰头宽 4cm。直裙结构如图 6-4-1 所示，前后片各有二个省道，后中开衩，后腰装拉链。

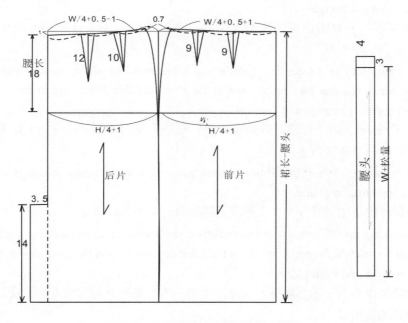

Fig. 6-4-1 Patterns of straight skirts
图 6-4-1 直裙结构图

1. Set size specifications / 规格设置

Open pattern-making system – click "New" – set size specifications or only select one base size, and directly enter the system to make patterns.

打开打板系统—单击"新建"图标—设置规格或只选择一个基准尺码，直接进入系统打板。

2. Make skirt patterns / 裙片制作

（1）Select "Zhizun pen" – double-click and "rectangular" appears – click a point as a start, – input

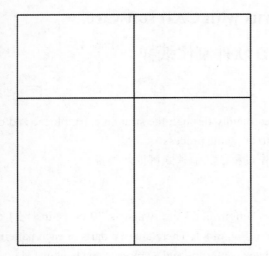

Fig. 6-4-2　The frame of straight skirt
图 6-4-2 直裙基本框架

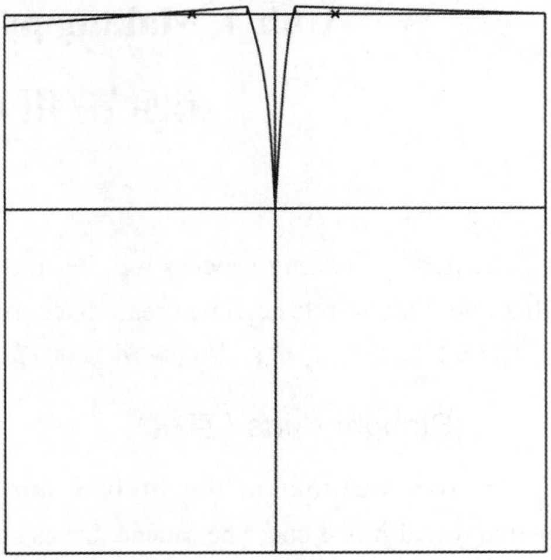

Fig. 6-4-3　Draw the back and front waist lines
图 6-4-3 画前、后腰围线

"47, 50" – enter. Get a rectangle.

选择"智尊笔"—双击出现"矩形"工具，在工作区单击一点作为起点—输入"47，50"—回车。得到一个矩形。

(2) Draw a parallel line 18 cm down to the top horizontal line of the rectangle and get the hip line. Divide the top horizontal line equally and make a vertical line downward at the bisection point. Get the frame of the skirt as shown in Fig. 6-4-2.

作距离上平线 18cm 的平行线，得到臀围线。平分上平线并在平分点向下作垂线。得到裙片基本框架图，如图 6-4-2 所示。

(3)Divide the waist width of the front and back pattern pieces into three equal parts and measure each part on the top horizontal line.

分别将前后片的腰宽三等分并在上平线上测量确定每个等分部分。

(4)Make cocking up 0.7 cm at the side seam of the waist line. Connect the cocking up point separateely to the point which drop 1cm along the back centerline, and the front center top point. Get waist lines of the back and front pieces.

在靠近侧缝处作 0.7cm 的起翘量。连接后中向下 1cm 处的点与起翘点，连接前中点与起翘点，得前、后片的腰围线。

(5) Make the side seam curve, as shown in Fig. 6-4-3.

作侧缝弧线，如图 6-4-3 所示。

(6)Divide the back waist line into three equal parts and separately make 12 cm, 10 cm long vertical lines downward – move 1/2 of dart width parallelly towards the right. In the same way do the darts of the front waist – then dig darts using "Zhizun pen". As shown in Fig. 6-4-4.

将后腰围线三等分，并分别作 12、10cm 长的垂直线—平行向右移动后腰省宽一半的量；同理作前腰围线上省。分别利用智尊笔挖腰省，如图 6-4-4 所示。

(7) Make a vent at the back centerline, 14 cm long and 3.5 cm wide.

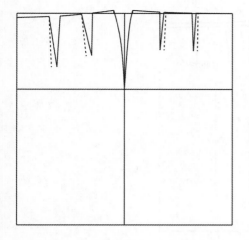

Fig. 6-4-4 Make waist darts
图 6-4-4 作腰省

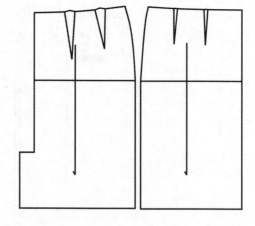

Fig. 6-4-5 Take out patterns
图 6-4-5 取出前、后片及腰头样板

作后开衩，长 14cm、宽 3.5cm。

(8) Take out the back and front pieces. Select "dart" in "handle pattern" – select "edit dart" – select "dart hill" – add a dart hill to each dart in the back and front pieces.

取出后片、前片，在"样片处理"菜单中选择"省"—选择"省编辑"—选择"省山"—将前、后片的省道加上省山。

(9) Make a waistband. Make a rectangle: width=W+2cm, height=4cm, and plus overlap 3 cm. Then take out it into patterns.

作腰头。作矩形：宽度 =W+2cm，高度 =4cm，再加 3cm 叠门。然后取成样片。

(10) Get the patterns of the front and back pieces, and waistband as shown in Fig. 6-4-5.

得到如图 6-4-5 所示的直裙的前片、后片及腰头样板。

(11) Set the front piece to "Symmetrical". Click the "Handle pattern" menu – click "Symmetry" – click "open pattern symmetrically " – click the front center line; on the opening piece, right-click on the front center line – click "Symmetrical Piece Close".

将前片设置为"对称"。点击"样片处理"菜单—点击"对称"—点击"样片对称展开"—点击前中线；在展开的样片上，将鼠标放在前中线处点右键—点击"对称片闭合"。

(12) Add seam allowance. Click "Handle pattern" – click "seam allowance" – select the pattern of front, back and waistband – right-click – input "1 cm" – enter. Select the hem of the front and back patterns – right-click – input "4 cm" – enter. Get the seam allowance pattern of the front, back and waist band, as shown in Fig. 6-4-6.

加缝份。点"样片处理"—点"缝边"—选择前片、后片和腰头—右击—输入"1cm"—回车。框选前后片下摆—右击，输入"4cm"—回车。得如图 6-4-6 所示各样片毛样。

二、Blouses / 女衬衫

The blouse is shown in Fig. 6-2-1 in unit 2 of this chapter. It is a fitted style with gathers on the front placket and waist darts on the back part. It has a fold collar, one-piece sleeves with cuffs.

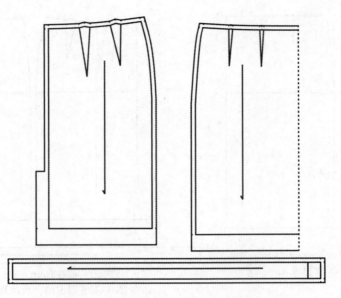

Fig. 6-4-6 Patterns of seam allowances
图 6-4-6 前、后片及腰头毛样

Drawing by prototype method. The size specification is as shown in table 6-4-1.

如本章第二节中图 6-2-1 所示女衬衫，款式比较合体，前片门襟部位抽碎褶，后片有腰省。翻领，一片袖，有袖克夫。采用原型法作图。规格如表 6-4-1 所示。

Table 6-4-1 The size specification   (unit: cm)
表 6-4-1 规格尺寸   （单位：cm）

Size/ 号型	Back center length / 后中长	Bust / 胸围	Waist / 腰围	Hip / 臀围	Sleeve length / 袖长
160/84A	58	94	80	94	58

1. Pattern making of the front and back pieces / 前、后片样板制作

See Fig. 6-4-7 / 见图 6-4-7。

(1) Take the prototypes of the front and back pattern pieces, and lenghten each of them by 20 cm. Adjust the collar hole, shoulder line and armhole.

取前、后片原型，将前后片分别加长 20cm。修改领围、肩线、袖窿。

(2) Make a dart in the back piece. Make the bust dart under the armpit and its dart width is the difference between the the front and back side seam lengths. Shorten each side of the waist line on the front and back by 2 cm.

作后片腰省。前片腋下作胸省，胸省宽为前后侧缝长之差。前片、后片侧缝收腰 2cm。

(3) Raise 4 cm at the side seam of the front and back hem, and draw hem curves.

前片、后片下摆侧缝处抬高 4cm，作下摆弧线。

(4) Take out patterns of the back and front pieces, as shown in Fig. 6-4-8.

取出后片、前片，如图 6-4-8 所示。

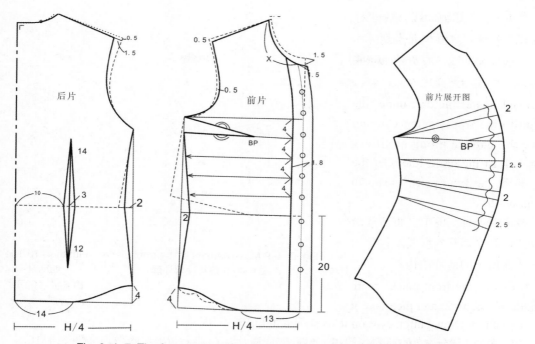

Fig. 6-4-7 The front and back structures of the blouse with gathers on chest
图 6-4-7 胸前褶皱女衬衫前、后片结构图

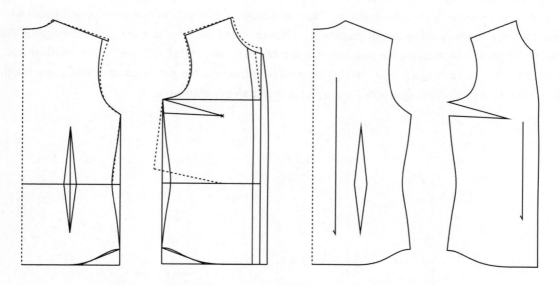

Fig. 6-4-8 The front and back patterns of blouse with gathers on chest
图 6-4-8 胸前褶皱女衬衫前、后片结构

(5) "Handle pattern" – "dart" – "definite dart" – select dart lines anticlockwise – right-click to end.
"样片处理"—"省"—"定义省"—逆时针方向选择省边—右击结束。

(6) Draw auxiliary lines with a space of 4 cm. Select "handle pattern" – "add elements into pattern" – select the auxiliary lines – right-click to end.

以 4cm 间距作辅助线。选择"样片处理"—选择"添加元素到样片内"—选择辅助线—右击结束。完成将辅助线添加到样片内部。

(7) Transfer the dart under the armpit to the front center. Cut and move the pattern along auxiliary lines. Adjust the grain line. Get the patterns of the front piece as shown in Fig. 6-4-9.

转移腋下省到前中。将前片沿辅助线剪开。调整丝缕线。得到如图 6-4-9 所示的前片结构。

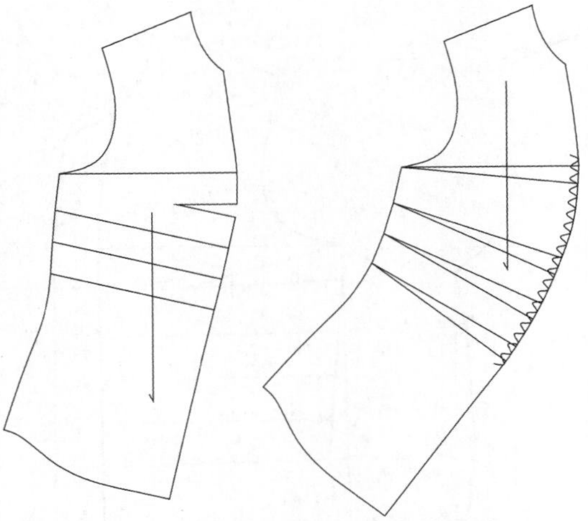

(8) Draw the front placket with a width of 1.8cm×2, and then take it out into pattern. Draw button eyes on it, as shown in Fig. 6-4-10.

绘制宽为 1.8cm×2 的前衣片门襟，然后取成样片。在其上画出纽眼位，如图 6-4-10。

Fig. 6-4-9 Make gathers of the front piece
图 6-4-9 前片抽褶处理

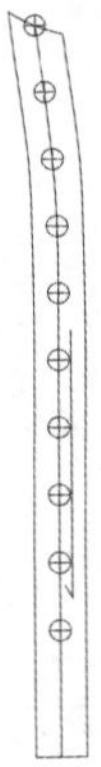

Fig. 6-4-10 The placket
图 6-4-10 门襟

2. Pattern-making of sleeve / 袖片制作

Reference to Fig. 6-4-11/ 参照图 6-4-11。

(1) First use the "curve measurement " tool to measure the front armhole curve length (front AH) and the back armhole curve length (back AH). Then draw the sleeve center line, and its length is the sleeve length (58 cm) minus the sleeve cuff width (6 cm) equals to 52 cm; useing the "vertical line" tool, take the sleeve crown height (1/3 of the total AH) from the top point and draw the sleeve width line, take the half sleeve length plus 2.5 cm and draw the sleeve elbow line .

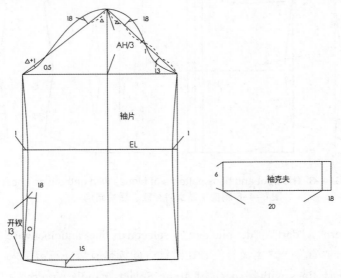

Fig. 6-4-11 The structure of sleeves and other parts
图 6-4-11 袖片及其他部件结构图

先用"测量弧线长度"工具，分别测量前袖窿弧线长（前 AH）、后袖窿弧线长（后 AH）。再画袖中线，其长为袖长（58cm）- 袖克夫宽（6cm）=52cm。再用"垂直线"工具，距顶点取袖山高（为总 AH 的 1/3），画袖宽线；取长为袖长 /2+2.5cm，画袖肘线。

(2) Use the "projection " tool to make slant lines of the front and back sleeve crown separately; the slant line length of the front sleeve crown is the front AH, and the slant line length of the back sleeve crown is the back AH plus 0.5 cm, and the two slant lines intersect separately with the sleeve width line.

利用"投影点"工具分别作前、后袖山斜线；使前袖山斜线长为前 AH，后袖山斜线长为后 AH+0.5cm，并与分别袖宽线相交。

(3) Draw the sleeve side seam line and sleeve hem line; Draw the sleeve side seam curve and sleeve hem curve.

作袖侧缝线直线和袖口直线；作袖侧弧线和袖口弧线。

(4) Draw the position of the sleeve slit, 13 cm high and 1.8 cm wide.

画出开袖衩的位置，高 13cm、宽 1.8cm。

(5)Divide the slant line of the front sleeve crown into four equal parts. Draw a vertical line at each of quarter points according to structure and so do the same way to the back sleeve crown slant line. Connect all points to get the sleeve crown curve, as shown in Fig. 6-4-12.

将前袖斜线四等分，在各等分点上作垂直线；后片类同。连接各点可得袖山弧线，如图 6-4-12 所示。

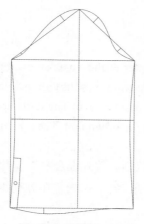

Fig. 6-4-12 The sleeve　　　　　　Fig. 6-4-13 The sleeve pattern
图 6-4-12 袖片　　　　　　　　　图 6-4-13 袖子样片

(6) Use the "Mirror" tool to turn over the sleeve vent position, and use the "Splice, correction" tool to align the bottom edge of the sleeve vent with the cuff curve, and then select the "Handle pattern - Take out pattern" tool to take the sleeve structure out into a pattern, as shown in Fig6-4-13.

用"镜像"工具将袖衩位置翻转，用"拼合修正"工具将袖衩底边与袖口弧线对齐接顺，再选择"样片处理—样片取出"工具，将袖子结构取成样片。如图 6-4-13。

1.Pattern-making of other parts / 其他部件的制作

(1) Cuff. Draw a rectangle of 20 cm by 12 cm. Take it out into a pattern.

袖克夫。作一个 20cm×12cm 的长方形，取出样片。

(2) Sleeve vent. Draw a rectangle of 13 cm by 3.6 cm. Take it out into a pattern.

袖衩。作一个 13cm×3.6cm 的长方形，取出样片。

(3) Collar. First draw the collar band part. As shown in Fig. 6-2-3, the length of the collar band is the sum of the front and back neckline curve lengths, the back width of the collar band is 3 cm, and its ends is cocking up 2 cm separately. Draw the structure of the collar band. Then draw the fall part. The back width of the fall is 5.5 cm, and the lower seam line of the fall part is 0.5 cm from the matching-up point of the upper seam line of the collar band. Draw the structure of the fall part.

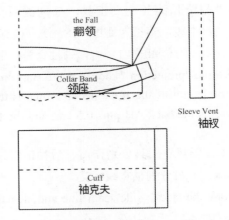

Fig. 6-4-14 Other parts
图 6-4-14 零部件

领子。先画领座部分。如图 6-2-3 所示，领座长取前、后领口弧线长之和，领座后宽为 3cm，其末端起翘为 2cm，画出立领部分结构。然后画翻领部分。翻领后宽为 5.5cm，翻领的下口线在领座上口线对位点处偏进 0.5cm。画出翻领部分结构。

(4) Finally you can get all parts as shown in Fig. 6-4-14.

最终得如图 6-4-14 所示各零部件。

三、Outer coats / 外套

The style shown in Fig. 6-3-4 in Unit 3 of this chapter is chosen as an example. It is a well fitted style, with a knife-back seam on the front piece and a transverse dividing line and a longitudinal dividing line on the back piece, lapel collar, two-piece fitted sleeve. The darts are hidden in the style lines. This pattern-making design is based on the prototype for women. The size specification is shown in table 6-4-2. The structure is as shown in Fig. 6-3-5.

这里选用本章第三节中图 6-3-4 所示款式作为范例。款式很合身，前片有刀背缝，后片上设有横向分割线和纵向分割线，翻领，两片合体袖子。省道被隐藏在分割线中。采用原型法作图。规格如表 6-4-2 所示。结构图如图 6-3-5 所示。

Table6-4-2 The size specification   (unit: cm)
表 6-4-2 规格尺寸 （单位：cm)

Size/ 号型	Back center length / 后中长	Bust / 胸围	Waist / 腰围	Shoulder width / 肩宽	Sleeve length / 袖长
160/84A	72	94	76	39	58

1. Pattern making of the front and back pieces / 前片、后片制图

(1) Copy the front and back bodice prototypes. The front and back body are separately lengthened 34 cm downward along the center line, and as shown in Fig. 6-3-5, draw the front and back frames.

复制前、后片衣身原型。前、后片衣身分别沿中心线处向下加长 34cm，按图 6-3-5 所示，作出前、后片框架图。

(2) The back neckline is widened by 1 cm and deepened by 0.5 cm, and adjust the back neck curve. Take half shoulder width plus 0.5 cm to get the shoulder point. Deepen the armhole depth by 1cm and adjust the armhole curve. Shorten the waist line from the side seam by 2cm. The hem at the side seam is widen by 5 cm and raised by 2 cm. Draw the hem curve.

后领口加宽 1cm，加深 0.5cm，重做后领围线；取肩宽 /2+0.5cm 得肩点，袖窿加深 1cm，重作袖窿弧线。侧缝收腰 2cm，下摆处加宽 5cm 并抬高 2cm，作下摆弧线。

(3) Widen the front necking by 1 cm, make the length of the front shoulder line shorter than that of the back shoulder line by 0.5 cm. Deepen the armhole depth by 1cm and adjust the armhole curve. Shorten the waist line from the side seam by 2cm. The hem at the side seam is widened by 5 cm and raised by 2 cm. Lengthen the front centerline by 5cm and then move to the left horizontally by 4cm. Draw the hem curve.

前领加宽 1cm，前肩线长度比后肩线短 0.5cm，袖窿加深 1cm，重作袖窿弧线。侧缝收腰 2cm，下摆处加宽 5cm 并抬高 2cm，前中线加长 5cm，水平向左 4cm，作下摆弧线。

(4) Get the frames of the back and front pieces and auxiliary lines as shown in Fig. 6-4-15.

得到如图 6-4-15 所示的前、后片框架及辅助线。

(5) Draw a horizontal cutting line 8cm below the nape point of the neck on the back center line. Take 1cm each side away from the midpoint of the waistline to make two longitudinal dividing lines. Make the cutting lines up or down 3 cm away from the waist line in the back center section, as shown in Fig. 6-4-16.

后中线上，后颈点下 8cm 处作横向分割线，在腰围线中点左右各取 1cm 处作 2 条纵向分割线，

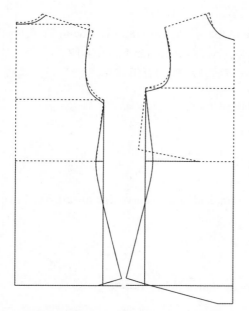
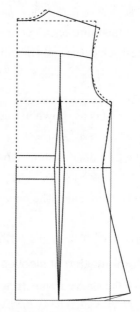

Fig. 6-4-15 The frames of the back and front pieces
图 6-4-15 前、后片框架

Fig. 6-4-16 Cutting lines on the back piece
图 6-4-16 后片分割线

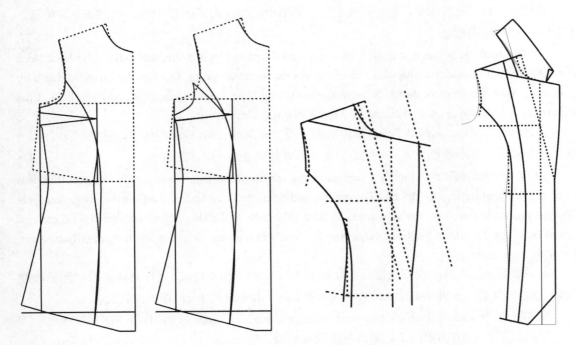

Fig. 6-4-17 The knife-back curve on the front
piece
图 6-4-17 前片刀背缝线

Fig. 6-4-18 The rever
图 6-4-18 驳头

Fig. 6-4-19 The collar
and facing
图 6-4-19 领片和挂面

后中片腰围线各向上向下 3cm 处作分割线。如图 6-4-16 所示。

(6) Make an underarm dart on the front, and its width is the difference between the back and front side seam lengths.

作前片腋下省，省宽为前、后片侧缝长度之差。

(7) Draw the knife-back curve on the front, and the waist dart width is 1.5cm. Transfer the dart into the knife-back curve and smooth the knife-back curve, as shown in Fig. 6-4-17.

在前片作刀背缝线，腰省宽为 1.5cm。将腋下省转移到刀背缝线中，修正刀背缝线，如图 6-4-17 所示。

(8) Make revers as shown in Fig. 6-4-18.

如图 6-4-18 所示绘制驳头。

(9) Make the collar as shown in Fig. 6-4-19.

如图 6-4-19 绘制领片。

(10) Make the facing. The upper end width is 3cm, and draw a curve downward, as shown in Fig. 6-4-19.

作挂面。上端宽 3cm，向下作弧线，如图 6-4-19 所示。

2. Pattern-making of sleeves / 袖片制图

(1) Make the sleeve frame using "Zhizun pen", as shown in Fig. 6-4-20.

作袖片框架，如图 6-4-20 所示。

(2) Draw vertical bisection lines of half the sleeve width line separately on both side of the sleeve

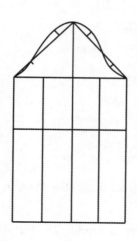

Fig. 6-4-20 The sleeve frame
图 6-4-20 袖片框架

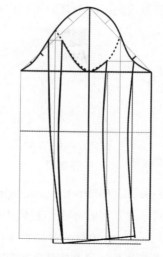

Fig. 6-4-21 The top and under sleeves
图 6-4-21 大、小袖片

midline and extend them to the sleeve crown curve. It is symmetric that the sleeve crown curves are on both sides of the bisection line.

在袖中线两侧分别作袖宽线一半的平分线，并向上延长到袖山弧线。在平分线两侧的袖山弧线是对称的。

(3) Make the front side seam line of top and under sleeves. Draw the side seam curves recessing 1 cm at the elbow line and extending up to the sleeve crown curve.

作大、小袖片的前侧缝直线。作侧缝弧线，前袖肘线处内凹 1cm，并向上延长至袖山弧线。

(4) Based on the bisection line of the front piece, draw a 12 cm long slant line from the point 0.5 cm up from the sleeve opening. Make the back side seam curve of top and under sleeve pieces, and extend it up to the sleeve crown curve.

以前袖平分线为基准线，从袖口向上 0.5cm 处作 12cm 的斜线，作大、小袖片的后侧缝弧线，并向上延长至袖山弧线。

(5) The top and under sleeves are as shown in Fig. 6-4-21.

图 6-4-21 所示为大、小袖片。

(6) Finally, take out all the patterns from the structure and name the pattern, fabric, and finish the pattern-making of the outer coat.

最后将衣片各部件从结构图中取出，填入样片名称及布料种类，完成外套制作。

(7) Add seam allowances to each component sample, and deal with seam corners to complete the cutting samples of all components.

将各部件样片加缝份，并处理缝份角，完成所有部件的裁剪样板。

Exercise/ 练习

1.Make the skirt pattern based on the front and back drawings of the skirt style below （Size：160/68A）.

根据下面裙子款式的正面和背面图制作裙子样板（号型：160/68A）。

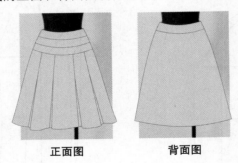

正面图　　　　　　　背面图

2.Choose two pieces of shirts to make their patterns according to the styles in the book.

根据书中款式选两款衬衫并制作它们的样板。

3.Choose two pieces of outerwear to make their patterns according to the styles in this book.

根据书中款式选两款外套并制作它们的样板。

Reference / 参考文献

[1] 章永红 . 女装结构设计（上）[M]. 杭州：浙江大学出版社 ,2005.

[2] 吴永红 . 服装结构设计 [M]. 重庆：西南师范大学出版社 ,2011.

[3] 张文斌 . 服装制版 基础篇 [M]. 上海：东华大学出版社，2012.

[4] 张文斌 . 服装制版 提高篇 [M]. 上海：东华大学出版社，2014.

[5] 张文斌 . 瑰丽的软性雕塑 [M]. 上海：上海科学技术出版社，2007.

[6] 竺梅芳 . GERBER 服装 CAD 使用教程 [M]. 上海：东华大学出版社，2006.

[7] 闫玉秀 . 女装结构设计（下）[M]. 杭州：浙江大学出版社 ,2012.